Mount Savage, Iron Empire

Patrick H. Stakem

(c) 2017

1st edition
Number 8 in the Western Maryland Series

To the Men and Women of the Advanced Manufacturing Center at Mount Savage, who contributed greatly to the success of the First Industrial Revolution in America.

- Author..4
- Introduction...4
- Mount Savage...6
- Timeline..7
  - The Companies 8
  - Captains of Industry..15
- Iron Production in Wales..18
  - Making pig iron...19
  - The raw materials 21
  - Flux..25
  - The workers...27
  - Mount Savage Blast Furnaces..28
- Production..31
  - Wrought Iron...33
  - Rolling the rails...33
  - Steel..39
- Mount Savage Railroad..39
- Cumberland & Pennsylvania Railroad...44
  - Locomotive Mount Savage...47
  - Steamship Mount Savage...47
  - Opening of the Mount Savage Railroad Extended..................49
- Locomotive Shops...52
  - Locomotive manufacturing..52
    - Engine Rebuildings at Mount Savage Shops........................57
    - Engines built at Mount Savage for other roads 59
- Contributions to Canal Boats..64
- The Curious Case of the Boiler in the Basement; a Mount Savage Thriller. 64
- Wrap-up..68
- Glossary..69
- Bibliography..71
- Resources..77
- Patents..79
- Glossary..81
- My WM/History books 83

Author

The author grew up in Cumberland, MD, across the street from the location of Fort Cumberland. He walked by it twice a day for his 8 years of elementary school, and decided to learn more about its history and reason for being. There was not much information available.

Mr. Stakem is a member of the Council of the Alleghenies, as well as the Western Maryland Chapter of the National Railway Historical Society, the Mount Savage Historical Society, the Westernport Heritage Society, and the C&O Canal Association. He currently resides in Laurel, MD. He has published numerous articles and books on the railroads, canal, and industrial and transportation heritage of Western Maryland. He is associated with Loyola University in Maryland and the Johns Hopkins University.

Special thanks for the help I received in collecting this material together: Allegany College in Maryland Library, Appalachian Collection; Bob Bantz; Allan Powell; "Champ" Zumbrun, expert on trees; Thomas Conlon; Al Feldstein; Ansel Shircliff; Richard Ibbotson of Sheffield, UK; Rev. Edward Chapman of Emmanuel Church; Lannie Dietle; the Preservation Society of Allegany County; Dr. Sarah Fatherly and many others.

Introduction

This book focuses on the early production of iron, and the later production of derivative products such as locomotives at Mount Savage, Maryland. Mount Savage was at the forefront of the Industrial Revolution in the United States. What made this possible was the juxtaposition of natural resources (iron ore, coal, and fire clay), availability of a workforce, and the relative ease of transportation to the coast. Nearby Cumberland had the B&O railroad from Baltimore, and the C&O Canal from Georgetown. The forward-seeing industrialists built their Mount Savage Rail Road not just to supply coal for shipment east, but for their iron products as well. This started with pig iron for the foundries in Baltimore, Georgetown, and along the Potomac. Significantly, the new short line railroad was built with rail they built themselves. THIS WAS A BIG DEAL. The B&O was using imported British Rail up to that point. Then this little upstart facility in the backwoods of Western Maryland came up with a cheaper, better local product.

Starting out repairing the *Baltimore Engines* from Ross Winans, the Mount Savage facility would go on to build maybe a hundred locomotives, for their own use, and for sale to other railroads in the U.S. and overseas. They built the coal hoppers as well, replacing Winans' pot hoppers with gondolas.

If we look at the importance of the Mount Savage facility in terms of the First and Second Industrial Revolutions in America, we can only compare it to Silicon Valley's influence on modern semiconductor technology.

The first Industrial Revolution started off in Great Britain. Iron making technology on an industrial scale was developed in Wales. Better iron lead to steam engines to pump out the water from coal mines. More coal lead to transportation by steam engine, and stationary steam engines replacing water for powering factories. This technology found its way to America, and was refined and improved. We had more things, and they were cheaper, because it was cheaper to manufacture and ship them.

Author

Mr. Stakem is the Historian, Western Maryland Chapter, National Railway Historical Society, Cumberland, MD, and a member of the Mt. Savage Historical Society, Council of the Alleghenies, Preservation Society of Allegany County (MD), C&O Canal Historical Society, Mountain State Railroad & Logging Historical Association (WV), Western Maryland Railway Historical Society (Union Bridge, MD). He is from Cumberland, Md. He has researched and written extensively on the Mount Savage Locomotive Shops, the furnaces, and the C&P Railroad. Two dollars from the sales of each book will be donated to the Mount Savage Historical Society.

Mount Savage

Mount Savage is a community in Allegany County with a current population around 873. Drive through the Town of Mount Savage today, and you would probably only remember it for the dog-leg turn on the narrow main street. You might miss the Iron Furnaces and the Castle on the right, the brickyard and the railroad buildings on the left.

It was in 1844 that Mount Savage was put on the nation's map with the rolling of the first iron rail in the United States. After this claim to fame, Mount Savage became the fifth largest city in Maryland. Today, no more iron is made at Mount Savage, nor do locomotives roll out of the shops. Little coal is mined, but the fire brick and refractory materials industry continue today.

Alternate (unofficial) names for Mount Savage have been Arnold Settlement, Corriganville, Jennings, Jennings Run, Jennons (sic) Run, Jennings Post Office, and Lulworth.

Many buildings in Mount Savage are on the National Register of Historic Places. The Mount Savage Historic District comprises 189 buildings, structures, and sites within the 19th and 20th century industrial community. The resources within the district reflect the community's development as a center of the iron, coal, brick, and railroad industries from the 1830s to the early 20th century. A broad variety of domestic, commercial, religious, and industrial buildings and structures represent all phases of the town's development during this period.

The town's commercial center is located along Main Street, and consists primarily of two and three-story commercial buildings dating from the turn of the 20th century. Most are of frame construction, but some are built with glazed brick, an architectural novelty produced in local brick works. A rich collection of domestic architecture is concentrated to the north, east, and southwest of the commercial area. Most of the houses are 1 1/2 or 2-story frame buildings, simplified interpretations of popular turn-of-the-20th-century styles, such as the Bungalow-influenced houses which line New Row and Foundry Row.

Late-19th century fashions are represented by notable frame Gothic houses, an Eastlake-influenced brick example, and a group of large frame Queen Anne houses. Several vertical-board duplexes overlook the former site of the Maryland and New York Iron and Coal Company operations, established in 1839. This site is currently occupied by the MOUNT SAVAGE Refractories brick works, the present descendant of the fire-brick industry which has operated continuously in town since the mid 19th century.

The Mount Savage Historic District is significant for its association with the industrial development of the Western Maryland region, and for its rich architectural resources representing a wide variety of types and styles of domestic, commercial, religious, and industrial buildings and structures reflecting all phases of the community's development from the mid 19th to the early 20th centuries. The vertical-board duplexes on Old Row are especially noteworthy as possibly the earliest examples of workers' housing remaining in the region.

Timeline

1608   Captain John Smith finds iron ore along the Patabsco River in the Colony of Maryland, sends two barrels to England.
1828   C&O Canal and B&O Railroad projects kick off on the same day.
1837   Blast furnace in Lonaconing in operation.
1838   Md. & NY Iron & Coal chartered in Maryland.
1839   Fire Clay mining in Mt. Savage; C&O Canal reaches Hancock.
1840-41 Mount Savage blast furnishes built; pig iron produced
1842   B&O Railroad reaches Cumberland.
1842   American RR Journal: "No firm in U. S. is capable of producing rail."
1842-47 Canal construction suspended.
1843   Rolling Mill set up at Mt. Savage.
1844   Rail production at Mt. Savage; iron ore discovered in Michigan.
1845   New York Mining Co.; Mount Savage Rail Road reaches Cumberland.
1846   Iron tariffs revoked
1847   Lulworth Iron Co. Mt. Savage; Detmold's tram road from

1848   Lonaconing to Clarysville.
1848   Mount Savage Iron Co.; largest producer in United States.
1850   C&O Canal reaches Cumberland; Mount Savage furnaces in full production.
1855   last iron production in Lonaconing.
1858   Bessemer Process for steel perfected in Pittsburg.

The Companies

This section discusses the various 19th Century companies establishing the iron business in Mount Savage.

Maryland & New York Iron & Coal Company (1838-1847)

On March 12, 1838, The Maryland and New York Iron & Coal Company was incorporated by Louis Howell, Benjamin B. Howell, and Henry Howell. The Howell Brothers of New York were bankers and brokers who could arrange for money for projects with a large anticipated rate of return. Benjamin Howell visited the Mount Savage area sometime before 1839, and liked what he saw. He ventured to England to gather Capital for an Iron Works; he succeeded to the extent of $600,000. The company was authorized to build a railroad from its mines at Mount Savage to the C&O Canal and the B&O Railroad at Cumberland by the terms of the charter: Both the B&O Railroad and the C&O Canal were on their way to Cumberland at this time.

Howell probably read Alexander's reports on the Lonaconing Project, picked out the ideas that worked, and discarded those that didn't. He build the Mount Savage facility in an area with the right raw materials (iron, limestone, and fireclay), but also with an existing community of potential workers, and infrastructure. Also, a rail connection to Cumberland was considered from Day 1. Howell purchased 3,700 prime acres from Andrew Bruce for $33,000. 1839 saw the construction of the first furnace. Until the railroad was completed to Cumberland, the right-of-way was used as a road, for horse-drawn wagons.

Louis Howell had vast land holdings and mines in Allegany County. The State charter allowed the Company to build or acquire railroads, as long as they did not interfere with the B&O, or the C&O Canal. They were also required to erect an Iron Works, and produce 1000 tons of pig, cast, or bar iron in one year. It is important to remember that when the legislature granted the rights to build a railroad, they included the right to condemn and acquire private land if it were needed to build the line.

Other investors included Joseph Weld, Thomas Weld Blundell, John Folliott Powell, Robert Samuel Palmer of the U.K., and several other American investors.

Louis Howell was lost at sea on the side wheel steamship *President* in 1841. The ship was at the time the largest steamship ever built. She was on her third crossing of the Atlantic. One could speculate that Mr. Howell was heading to England to raise Capital for his venture.

The State charter says:

"And be it enacted, That for the purpose of enabling said company to transport the produce of the mines and of the counties through which their rail road shall pass, on the cheapest and most expeditious manner, the said company and the president and directors thereof shall be, and hereby are respectively invested with all and singular the rights, profits, powers, privileges, authorities, immunities and advantages fur the surveying, locating, establishing and constructing a rail road and its necessary appurtenances, beginning the same at the mines of the said company and running to a convenient point or points on the basin or canal of the Chesapeake and Ohio Canal Company, at or near the town of Cumberland, in this State, and for the using, preserving and controlling the said rail road, its necessary vehicles and appurtenances and every part thereof, or borrowing money on the credit of the company for its lawful purposes; provided, that no such borrowing of money shall imply a right to borrow or purchase the stocks of the State, or any other description of property whatever, which by the act, and more particularly the fifteenth section thereof, incorporating the Baltimore and Ohio Rail Road Company, and its several supplements, were for the lawful purposes of said company, and the benefit of its corporators given, granted, authorized and secured to the said company and to the president and directors respectively, as fully and perfectly as if the same were herein repeated; provided, that" it shall not be lawful for the said Maryland and New York Iron and Coal Company to occupy or use any portion of the lands that may be necessary for the accommodation of the canal and works of the Chesapeake and Ohio Canal Company, or for the main route of the Baltimore and Ohio Rail Road, or that may be within the limits of either of the public roads there now existing, except to cross these roads without injury to the same; and provided also, that full right and privilege is hereby reserved to the citizens of this State, or any company now or hereafter to be incorporated under the authority of this State, to connect with the rail road hereby provided for, or any other rail road, if in the opinion and judgment of the commissioners of Allegany county, for the time being, passed upon hearing of all parties interested, no injury would be done by such connection to the rail road of said company, and that the said company shall transport on the said rail road at the rate of one cent a ton per mile on all goods, merchandise or property of any description whatsoever transported on said rail road, or on any lateral way which they may construct, and also not exceeding two cents per mile for each passenger transported on said road; provided always, that when any car shall be placed on said rail road it be adopted in size, and all necessary particulars to said rail road; and provided further, that the Legislature of this State may at any time hereafter regulate, modify or change the control, use, and estate of said rail road as shall be constructed under the authority hereby given, in such manner as it may deem equitable towards the said company, and necessary to the accommodation of the public travel or use of the said rail road." Later, in 1841, the Charter was amended:

"Be it enacted by the General Assembly of Maryland, That it shall be lawful for the Maryland and New York Iron and Coal Company, to charge, demand and receive, for all persons and property transported on the rail road and any lateral way, which they are authorized to construct from their mines to the basin of the Chesapeake and Ohio Canal, or other points, in or near the town of Cumberland, the same rates of toll, or prices of transportation as the Baltimore and Ohio Rail Road are, or shall be, by law allowed to charge and receive. And whereas, doubts may exist whether the said company would be authorized, under the act to which this is a supplement, to construct a rail road from their mines or Works, to some intermediate point or points between the basin of the Chesapeake and Ohio Canal at Cumberland and their works or mines aforesaid, should circumstances render the extension of their road to the basin of the canal at Cumberland unnecessary, as will probably be the case, if the Baltimore and Ohio rail road company, or the said canal company, extend their works up the valley of Jennings' run, to give additional facilities to the coal and iron trade—therefore,"

"Be it further enacted, That it shall not be necessary for the said Maryland and New York Iron and Coal Company, to construct their said rail road or any lateral way from their said works or mines to the basin of the said canal at Cumberland, but that the same may be stopped at any intermediate point, at the discretion of the company, and that it shall be lawful to charge the same rates of toll for the transportation of persons and property upon such road when constructed, as are authorized by the first section hereof; provided, that the said rail road be constructed so as to intersect with the extension of the Baltimore and Ohio rail road, or the Chesapeake and Ohio canal, or the improved navigation of Wills' creek, by canal or otherwise."

"And be it further enacted, That a quorum for the transaction of business of the said Maryland and New York Iron and Coal Company, shall hereafter consist of the President and any two Directors, as required by the fourth section of the act to which this is a supplement "

The Mount Savage rolling mill was built in 1843 by the Company, but it was not the path to financial success that they hoped. The company had to borrow $30,000 from Mr. Semmes, but that was only a short-term proposition. In 1848, the company failed, partially because Congress had decided to lift protective tariffs on British rail. The property was sold at auction to John M. Forbes of Boston, who conveyed it to the Lulworth Iron Company, which had been incorporated in 1846.

Lulworth Iron Company (1847-1848)

The Lulworth Iron Company was chartered in the state of Maryland on March 1, 1847. The key players were Samuel M. Semmes, John G. Lynn,

Henry Thomas Weld, Jonathan Guest, and Robert Samuel Palmer. They were empowered for "...carrying on the manufacturing of iron, and of articles of which iron is a component part, and for opening, working, transporting to market and vending the products of their lands, mines, manufactories..."

The charter continues, "That for the purpose of enabling said corporation to transport the produce of its mines and manufactories to market and elsewhere, in the cheapest and most expeditious manner, the said corporation and the president and directors thereof, shall be, and are respectively invested with all and singular the rights, profits, powers, privileges, authorities, immunities and advantages for the surveying, locating, establishing and constructing a rail road or rail roads, with the necessary appurtenances, beginning the same at or near the mines or manufactories of the said corporation, and running to a convenient point or points at or near the town of Cumberland, or to such other point or points as may best suit the convenience and interest of said corporation, and for the using, preserving and controlling the said rail road or rail roads, and the necessary vehicles and appurtenances thereto belonging, and every part thereof, which by the act, and more particularly the fifteenth section thereof, incorporating the Baltimore and Ohio Rail Road Company and its several supplements, were for the lawful purposes of said company, and the benefit of its corporators given, granted, authorized and secured to the said company, and to the president and directors respectively, as fully and perfectly as if the same were herein repeated; provided, that it shall not be lawful for the said Lulworth Iron Company to occupy or use any portion of the lands that may be necessary for the accommodation of the canal and works of the Chesapeake and Ohio Canal Company, or for the main route of the Baltimore and Ohio Rail Road, or that may be within the limits of either of the public roads there now existing, except to cross these roads without injury to the same; and provided also, that full right and privilege is hereby reserved to the citizens of this State, or any company now or hereafter to be incorporated under the authority of this State, to connect with the rail road or rail roads hereby provided for, or any other rail road, if in the opinion and judgment of the commissioners of Allegany county, for the time being, passed upon hearing of all parties interested, no injury would be done by such connection, to the rail road of said corporation; and that the said corporation shall transport on its said rail road or rail roads, all persons and properly, at the same rates of toll and prices of transportation as the Baltimore and Ohio Rail Road Company are, or shall be, by law, allowed to charge and receive; provided however, that in all cases where a connection is formed between the rail road or rail roads hereby authorized to be constructed, and the rail road or rail roads of any other corporation or citizen of this State, the cars to be used in the transportation of persons and property shall be adapted in size and all necessary particulars to the rail road or rail roads of the said Lulworth Iron Company; and provided further, that the Legislature of this State may at any time hereafter regulate, modify or change the control, use and estate of the rail road or rail roads to be constructed under the authority hereby given, in such manner as it may deem equitable towards the said corporation, and necessary to the accommodation of the public travel or use of the said rail road or rail roads."

In the early deeds and records the community around the furnaces was called Lulworth because the Lulworth Iron Company once owned the clay and manufacturing rights. But later the name Mount Savage appeared when it was sold again, taking its name from the one the people living in the area preferred. As a matter of record, however, the community that was slowly growing up around the brick yard, blast furnaces and railroads, was called "Savage Mountain Hamlet", but as the town grew larger, the Hamlet was dropped and "Savage Mount" continued in use for many years. Whether the "Mount Savage" had a more lyrical sound than "Savage Mount" is not known, but it became reversed and ever since was called Mount Savage. Lulworth Iron later changed its name to Mount Savage Iron Company on Feb. 7, 1848.

Mount Savage Iron Company (1848-1867)

A major investor in Mount Savage Iron was Erasmus Corning, of New York. He made John F. Winslow the President of the Company. The Mount Savage Rail Road had been built with "Winslow's Patent Rail." This was a British patent.

John A. Graham (the first president of the Cumberland & Pennsylvania

Railroad) with fellow directors John F. Winslow, Warren Delano, John M. Forbes, and Joseph B. Varnum ran the company.

Lulworth Iron was essentially split into two parts. The railroad went to the C & P, and the iron manufacturing went to Mount Savage Iron. The two companies shared directors.

Mount Savage Iron did not pay cash for the shares; rather, it conveyed its existing railroad operations, stretching from the B&O depot in Cumberland and the Potomac Wharf to the mines near Frostburg, to the newly formed C & P Railroad. Thus, Mount Savage Iron was out of the railroad business directly, and the C & P was in. Mount Savage Iron completed the Canal Wharf (ex-Lynn Wharf) in Cumberland in 1850. Canal boats were loaded in the Potomac, then entered the canal via the guard locks. Mount Savage Iron operated the Mount Savage Rail Road until 1854, when it went under C & P control.

New York Mining Company (1845)

The New York Mining Company was chartered in the State of Maryland in February 26, 1845. The incorporators were Oroondates Mauran, Barrett Ames, Robert B. Minturn, Jonathan Sturges, Charles Dennison, and Samuel M. Semmes. It was allowed to mine coal and iron, and manufacture and sell iron products. Mr. Mauran was a wealthy New York businessman who, with his partner Cornelius Vanderbilt, owned the Staten Island Ferry. Robert Bowne Minturn was one of the most prominent American merchants and shippers of the mid-19th century. Today, he is probably best known as being one of the owners of the famous Clipper Ship, the *Flying Cloud*. He was a New York merchant, involved in the China and transatlantic trade. He and his wife donated the land for New York's Central Park.

Mount Savage Fire Brick Company (1841-present)

The Mount Savage Brick Works established its first plant in 1841. The bricks were shipped all over the country. The Mount Savage Firebrick Co., now located in Zihlman, is still in operation, and its products are used for specialty steel mills.

Captains of Industry

This section discusses some of the movers and shakers, the enablers and the technology experts who came together to make Mount Savage Industry happen.

Samuel Middleton Semmes
Semmes was born in Charles County, Maryland, in 1811. His brother Raphael went on to become an Admiral in the Confederate States Navy. Samuel graduated from Georgetown Law, and was admitted to the bar in Allegany County. He drafted most of the charters of the pioneering coal companies in Allegany County. He served as State Senator from 1855 to 1866. He was associated with Lulworth Iron, and the New York Mining Company. Semmes died in 1867.

John Galloway Lynn
The Lynn's were an 18th century Maryland family. J. G. Lynn's Father had moved to Cumberland, building a substantial brick house and estate on the West Side known as Rose Hill. John Galloway became a Cumberland businessman, and built the Lynn Wharf along the Potomac River. This was sold by his heirs to the Maryland Mining Company. In 1849, Lynn and the Mount Savage Rail Road incorporated the Cumberland and Pittsburg (sic) Rail Road Company. Their eyes were on the grade over the Alleghenies to Pittsburgh, but nothing seems to have come of the venture.

Henry Thomas Weld
Weld was an English immigrant. The Weld's Lulworth Estate is located in central south Dorset, England. Its most notable landscape feature include a five mile stretch of coastline on the Jurassic Coast. Part of the area is now a special World Heritage Site. The estate is predominantly owned by the Weld family who have lived there for several generations. The Lulworth estate was once part of a grand estate under Thomas Howard, 3rd Viscount Howard of Bindon. The historic estate, hosted the stately Lulworth Castle, which was the residence to the Weld family until 1929 when it was ravaged by fire. Henry was associated with Welds Boatyard in Cumberland, later the Weld & Sheridan Boat Building & Repair Yard at

the C&O Canal basin.

Warren Delano
Warren Delano II was the maternal grandfather of President Franklin Delano Roosevelt. During a period of twelve years in China, Delano made more than a million dollars in the tea trade in Macau, Canton, and Hong Kong, but upon returning to the United States, he lost it all in the Panic of 1857. In 1860, he returned to China and made a fortune in the controversial but highly profitable opium trade, supplying opium-based medication to the U. S. War Department during the Civil War. The Delano Mining Company operated in Mount Savage, producing coal. His partner was James Roosevelt around 1870. The future President spend some summers with his grandfather, who lived at the Bruce estate. Delano was a director of Consolidation Coal 1864-1875, and a Director of the C & P Railroad. He had a C & P locomotive named after him, a Winans Camel, in 1859.

John M. Forbes (1813-1898)
Along with Jay Gould and E. H. Harriman, John Murray Forbes of Boston was an important figure in the building of America's railroad system. From March 28, 1846 through 1855, he was president of the Michigan Central Railroad, and a Director and President of the Chicago, Burlington and Quincy Railroad. He spawned a vast 19th century Industrial Empire, starting early in the century in the China Opium trade. He and his partner Erasmus Corning of Albany bought the Mount Savage Iron Works for $200,000, which was about one fifth of its value. The facility had gotten into cash flow problems, and this was the kind of deal Corning looked for. He and Forbes added the rolling mill that produced the first successful iron rail in America. The "Forbes Group" went on to acquire railroads all across America.

In *Letters and Social Aims*, Ralph Waldo Emerson said of Forbes: "Never was such force, good meaning, good sense, good action, combined with such domestic lovely behavior, such modesty and persistent preference for others. Wherever he moved he was the benefactor. How little this man suspects, with his sympathy for men and his respect for lettered and scientific people, that he is not likely, in any company, to meet a man superior to himself," and "I think this is a good country that can bear such a creature as he." (Forbes' son married Emerson's daughter).

John F. Winslow
Winslow was a partner of Erasmus Corning, and worked at the Albany and the Rensselaer Iron Works in New York in 1837. He was an engineer, iron master, the inventor of compound rail, and President of the Mount Savage Iron Company in 1848. He traveled extensively in Europe in 1852, buying the rights to iron and steel processes. In 1861, he partnered with John Ericson on his Navy contract to build an iron clad war ship, the *Monitor*. Some of the techniques he developed for making hardened iron plate may have come from his work at Mount Savage in the period 1848-1852.

Besides a British patent for rail, Winslow held several American patents, including number 35407, 1862, for "Improved Armor Plate for Vessels," number 34177, 1862, for "Compressing Puddle Balls," and number 4526 of 1846 for "Malleable Iron from Ores." This latter was when he worked for Corning in Troy, NY. He was also a Director of the Cumberland & Pennsylvania Railroad.

William Borden
He was president of the Borden Mining Co., circa 1875. The Borden Family was from New York. They had founded the water-powered industrial town of Fall River, Massachusetts, outside of Boston.

Enoch Pratt (1808-1896)
Pratt was a Capitalist, and a friend of Andrew Carnegie. He was born in Massachusetts, where he learned iron making. He arrived in Baltimore in 1831 with $150, and went on to make his fortune. E. Pratt & Brothers (Hardware), 23-25 S. Charles St., Baltimore. He was a director of the Maryland Steamboat Co., Director of the Susquehanna Canal Co., Vice-President of the Philadelphia, Wilmington, & Baltimore Railroad, and director of three other Railroads. He built and donated a public library system to the City of Baltimore.

The following sections will trace the technology of iron manufacturing at Mount Savage back to the earlier furnace at Lonaconing, and that in turn back to the iron manufacturing centers at Wales.

Iron Production in Wales

One of the world's major iron producing regions for the Industrial Revolution was Wales. Most of what we know of Welsh iron making in the period is from the papers of the Tredegar Iron Works. (later, an iron works in Richmond, VA, would be named after this facility.) In Welsh, Tredegar means "ten acre town."It was noted for its abundant natural resources including iron ore and coal, plus water power. It's only problem was its remote location. The facility was built in the age of charcoal, and later transitioned to coke as a fuel. Incidentally, the Welsh word for forge is "Ffwrwm."

The Sirhowy iron works were established in Wales in 1750. It used a coal-fired blast furnace and a forge. This went on to become the Tredegar facility. Earlier forges were built in the area beginning in 1690, with facilities dating from the 1560's. The owners of the facility were Roger Powell, Roger Williams, and John Morgan. The facility produced pig iron, and bar iron, and was profitable. The facility shipped 150-200 tons of bar iron per year. A major customer was Foley's wire works in Tintern. Surviving records show that 2 ½ loads of charcoal and 2600 pounds of pig were needed to produce a ton of bar, in the forge. Initially, additional pig iron was also imported to feed the forges. Water-powered bellows provided the blast.

The town had an influx of workers, and needed housing. This was provided by entrepreneur Samuel Homfray. The population exploded from 1100 in 1801 to nearly 35,000 by 1881. An interesting perspective on the Town in 1832 is provided in an 1985 book by Adrian Vaught, "Grub, Water, & Relief."

"Utterly remote at the head of the Sirhowy valley, the town was a man-made hell. Men and children worked killing hours in the smoke and filth of the foundries and were maimed by molten metal. Their only medical help was that administered by the 'Penny Doctor.' Wages were paid in Homfray's private coinage — banks were not allowed in the town — so workers spent their coins in Homfray's shops, buying food at Homfray's prices. Poverty and malnutrition followed and disease followed both."

Merthyr Tydfil was once the largest town in Wales. Merthyr was close to reserves of iron ore, coal and limestone and to water transportation, making it an ideal site for ironworks. Small-scale iron working and coal mining had been carried out at some places in South Wales since the Tudor period, but in the wake of the Industrial Revolution the demand for iron led to the rapid expansion of Merthyr's iron operations. Neighboring Dowlais Ironworks was founded by what would become the Dowlais Iron Company in 1759, making it the first major works in the area. The demand for iron was fueled by the Royal Navy, which needed cannon for its ships, and later by the railways. Several railway companies established routes linking Merthyr with ports and other parts of Britain. They included the Brecon and Merthyr Railway, Vale of Neath Railway, Taff Vale Railway and Great Western Railway. They often shared routes to enable access to coal mines and ironworks through rugged country, which presented great engineering challenges. In 1804, the world's first railway steam locomotive, "The Iron Horse", developed by the Cornish engineer Richard Trevithick, pulled 10 tons of iron on the new Merthyr Tramroad from Penydarren to Abercynon.

Welsh Iron Masters and workers came to Lonaconing to make the furnace there a thriving business. The Lonaconing furnace was modeled on the Welsh *Cyfarthfa* furnace. The furnaces at Mount Savage were modeled on the Lonaconing facility. We should take a look, then, at the state-of-the art in Welsh iron working in the 1830's.

Making pig iron

Pig iron, so called because of the arrangement for tapping the furnace, is not terribly difficult to make. You need iron ore, a fluxing agent, a source of pure carbon, and enough heat to get to 2797 degrees F. The Welsh got the process working on an industrial scale by the 1600's.

Pig iron is the product of smelting iron ore with coke or charcoal using limestone as a flux. It has a very high carbon content, typically 3.5–4.5%, which makes it very brittle and not useful directly as a material except for limited applications.

The preferred fuel is coke, nearly pure carbon, made from coal. The preferred blast is heated air. The limestone serves as a flux, to collect the impurities from the ore. Iron ore was mined locally, and limestone came from nearby quarries. Coal was burned into coke on site in long pits. This removed the sulfur and phosphorous, which interfered with the iron extraction process. The process of extracting iron from ore is less of a melting process than a chemical reduction process. The carbon from the coke binds with the oxygen from the iron oxides in the ore, and goes off as carbon dioxide and carbon monoxide. Sometimes, the iron ore was also roasted before being introduced into the furnace. This served to remove contaminants present in the raw ore.

Closer to the coast, limestone was not abundant, but the iron facilities used ground sea shells. Same stuff. Calcium Carbonate.

The traditional shape of the molds used for these ingots was a branching structure formed in sand, with many individual ingots at right angles to a central channel or runner. Such a configuration is similar in appearance to a litter of piglets suckling on a sow. When the metal had cooled and hardened, the smaller ingots (the pigs) were simply broken from the much thinner runner (the sow), hence the name pig iron. As pig iron is intended for remelting, the uneven size of the ingots and inclusion of small amounts of sand was insignificant compared to the ease of casting and of handling.

Pig iron is then remelted and a strong current of air is directed over it while it is being stirred or agitated. This causes the dissolved impurities (such as silicon) to be thoroughly oxidized. The metal is then cast into molds or used in other processes. Good for plows, pots, and horse shoes.

The furnace can be operated continuously, as liquid iron and slag are drawn off at the bottom, then the tapping holes are re-plugged with clay. The furnace is "charged" or filled at the top, which is why many furnaces are built against a hillside. The continuous process is efficient, since the furnace doesn't need to cool be reheated. Generally, a furnace will last about a year, before it has to be rebuilt or relined.

In the simpler cases, a furnace is charged, allowed to burn out, and recharged. This would be done if there is no convenient way to get wheelbarrows of the raw material to the top. Both Lonaconing and Mount Savages were built against a hillside, for ease of "charging" the furnace.

The raw materials

The Mount Savage area has a unique combination of plentiful coal, limestone, iron ore, and fire clay, the ideal basis for a 19th century industrial base. Nothing is more than a few miles away. Given iron ore, limestone, coal, and fire clay for a furnace lining, you can certainly make pig iron.

Fire clay is a specific kind of clay used in the manufacture of ceramics, especially fire brick. The material is named for its refractory characteristics. Fire clay was discovered in the Pottsville formation near Mount Savage in 1837.

Products made from fire clay are resistant to high temperatures, and have a fusion point higher than 1,600°C. They are then suitable for lining furnaces, and as fire brick, and manufacture of utensils used in the metalworking industries, such as crucibles and glassware. Because of its stability during firing in the kiln, it can be used to make complex items of pottery such as piping. Its chemical composition consists of a high percentage of silicon and aluminum oxides, and a low percentage of the oxides of sodium, potassium, and calcium. Unlike conventional brick-making clay, fire clay is mined at depth, usually found underneath a coal seam. Very convenient. The formations in Allegany County range from 5 to 20 feet thick.

Mining of fire clay in Mount Savage began in 1839, and continues to this day. It was also brought from mines on Big Savage Mountain west of Frostburg. The local product was substituted for imported British product. It became a sideline business for the Maryland and New York Iron & Coal Company, but a very necessary one.

The Mount Savage mines were some 2 ½ miles west of the town. The clay bed was from 8-14 feet thick, with a thin layer of coal on top. It contained both hard and soft clay. By 1864, the Mount Savage fire clay became the standard in the United States.

In 1914, The Union Mining Company's four fireclay mines were located 4 miles west of Mount Savage, on Savage Mountain. S. J. Aldon was the Superintendent, with Joseph Jenkins as the Mine Foreman. The mine cars of clay were hauled to the surface by mules and dumped into larger cars that went down a long plane by gravity. The loaded cars going down the plane hauled the empty cars back up by a cable and pulley. At the bottom of the plane, a small ('dinky") locomotive hauled the cars two miles to the Mount Savage yards. In the year of 1914, these mines employed 54 men, and produced over 33,000 tons of fire clay.

The Savage Mountain Fire Brick Company had a similar operation, but used horses in place of the locomotive. The clay went by wagon to Mount Savage by way of Frostburg. Production was 10,780 tons in 1914.

The Big Savage Fire Brick Company used mules inside the mine, and a stationary engine to bring the cars a distance of 2 ½ miles to the brick yards at Allegany, on the Cumberland & Pennsylvania Railroad. They produced 11,880 tons per year.

The Mount Savage Refractory is still operating under the auspices of Mr. A. J. Rost of Pittsburgh, who purchased the assets of the Union Mining Company in 1944. This makes the Mount Savage facility the oldest continuously operating firebrick plant in the United States.

There are two major types of fire clay used in the production of bricks. Flint clay, or hard clay, exhibits little or no shrinkage when fired. Plastic, or soft clay absorbs water and is easily workable. The correct mixture is critical. Both types consist of silica and alumina, in different proportions. The deposits available to Mount Savage include both types of clay.

Captain John Smith is generally credited with the discovery of iron ore in Maryland in 1608. There is no evidence that the Native American population ever smelted ore, although they did use it as a pigment. George Washington's father was involved in the early colonial period iron industry. The iron ore in the vicinity of Mount Savage had a particularly useful yield. It is carboniferous and from the family of Clinton Ores, which are hematite (Fe_2O_3).

Iron ore was mined in various locations in the county, for the furnaces at Mount Savage. One mine that served the Mount Savage facility was on the Samuel Eckles property on the west side of Will's Mountain. This area was worked from 1845 to 1855, and re-opened during the Civil War. Iron ore was mined in the town of Mount Savage, on the north and west side. The larger mines were on the hill called Ridgeley, about 1 1/2 miles west of town, on the north side of Dutch Hollow. At the foot of the Mount Savage gravity plane, the "Lower Tunnel" was opened in 1846, and was worked until 1853. This tunnel was reported to be 1/2 mile long. The "Upper Tunnel" was located on the property of Henry Collins, and was also about 1/2 mile long. These excavations also yielded fire clay.

Iron melts at 2797 degrees F. Actually, it freezes at 2797 degrees F as well. But cast iron, or pig iron is at best an intermediate product. It is not strong in tension, and not suited for many applications beyond casting simple implements. The follow-on process makes wrought iron from pig. Small batches of wrought iron can be produced by a blacksmith at a forge. Larger production requires a different type of furnace.

Iron ore from George Jeffries & Sons nine in Frostburg went to Mount Savage. About 5,000 tons were provided at a cost of $4.50 to $5.00 per ton. The Jeffries paid a royalty of $.25 per ton to the Frosts, owners of the land. The ore bed was 18 inches thick, and was recovered by drift or strip mining. This mine closed about 1855. Jeffries then worked a mine during the Civil War on the Johnson property in Frostburg. This was a parcel of land 2 acres in extent, with a vein 4 feet thick. About 10,000 tons were extracted by 10 men, and numerous teams of horses. Joseph Johnson got a royalty of $.30 per ton.

Iron ore, author's collection.

Bituminous coal is abundant in the Mount Savage region, as part of the Big Vein formation. Coke is better for iron making, being more pure carbon with fewer impurities. The pioneering use of coke for iron making was the furnace at Lonaconing. Previously, charcoal had been used, leading to a massive destruction of trees to feed the iron furnaces.

Coal was used from the local mines which gave rise to a long period of prosperity in manufacturing bricks, mining coal, and building engines and cars at the railroad shops. Clay was shipped to other manufacturing towns that made cement, lime pottery and enamel ware.

Coke is made by burning coal in a closed atmosphere, which drives off the impurities, and makes the product a more pure carbon. It is a destructive distillation process. Impurities such as sulfur will interfere with the purity of the iron produced, essentially poisoning the batch. The coal gas can be captured and used. At Mount Savage, it was used for the reheating furnaces. It can also at the same time provide municipal gas for cooking and lighting, In addition, passing steam or just air over hot coke produces a very clean gas product.

In 1849, coke was produced in Mount Savage on a ledge dug out of the hill, on a level with the top of the iron furnace. The iron furnaces were loaded, or "charged" from the top. These were the first production coke ovens in Maryland. There were 28 of them, said to be smaller than the Connellsville (PA) ovens. At Mount Savage, the coke was taken out at "white heat," drenched with water, and left to cool. Coke ovens can reach 2000 degrees Celsius. The process takes a few days.

Coke is less dense than coal, thus easier to transport. Coke cars for railroad use can be larger than coal cars, without exceeding weight limits.

Limestone is quarried, not mined. There are many exposed outcroppings of the rock in Western Maryland. It does need to be crushed into smaller pieces. It can then be burned to lime for cement, or for use as a flux for the furnaces. The *Coal Measures* limestone is used for fluxing iron.

Flux

The fluxing agent binds to the impurities from the iron, and cools to a material called slag. Slag is usually glassy, and has different colors, depending on the impurities. It's mostly a waste material, but can be used for roadbeds. You can find a lot of it walking from the Mount Savage Post Office back to the furnaces.

Limestone for the Mount Savage furnaces was obtained from the Dunkart Formation, some 2 miles South-East of Frostburg, and from near Corriganville, where the limestone is some 30 feet thick. In fact, the Devil's Backbone formation with its nearly vertical strata, is all limestone. There are also limestone foundations near Barrellville.

In essence, iron ore is rust, iron oxide with some impurities. If we remove the bonded oxygen from the iron oxide, we get the pure iron back. That's the job of the carbon monoxide, produced from the burning coke, it becomes carbon dioxide. This is facilitated by high temperature. Thus, in the furnace, the iron ore is not melted – the oxygen is grabbed from it by the carbon. The free iron then melts and flows. The resulting slag is removed from the furnace, broken up, and used for road building, railroad ballast, and fill – nothing is wasted. The limestone also absorbs impurities from the raw iron, materials such as silicates, sulfur, and phosphorus. The Georges Creek coal is low in sulfur and phosphorus, which are referred to as "penalty elements." Most of what we find mixed with iron ore results in a less-than desirable batch of iron. Sulfur is bad, alumina (Al_2O_3) and magnesium oxide (MgO) can be good.

Glassy slag from Mount Savage, author's collection
slag, Mount Savage. Note striations.

Slag, Mount Savage

The workers

Many Irish, Welsh and English came here to work at the furnaces, rolling mills, brickyard, coal mines, and the railroad. The skilled ironworkers came from Wales, and the Irish were mostly common laborers. Competition for labor came from the ongoing railroad and canal construction projects, and drove up wages. The towns were cultural melting pots.

The Mount Savage Iron Works built 280 company houses for their workers in 1847. They were part of the deal, the social contract. The company also built a store and a church, and employed a company doctor, similar to Lonaconing.

The brickyard at Mount Savage was lucky to get one John Davis, a former Welsh rolling mill worker. He was of tremendous height and strength. His job was to straighten the crooked rails with a 100 pound mall. He was the only man who could lift and swing it. The mall is still at the brickyard and it is referred to as *The John Davis*. Davis is Mount Savage's "John Henry."

Mount Savage Blast Furnaces

The furnace at Lonaconing was a model for the ones at Mount Savage, and there was technology sharing, if not downright industrial espionage between the two facilities. Three furnaces were built at Mount Savage, but only two went into service. The two that did go into blast resembled the furnace at Lonaconing, fifty feet high, fifteen wide at the bosh, and built against the side of a hill. They were on the south side of Jennings run. The third furnace built in the open and would have had to be loaded by derrick. The furnaces were lined with firebrick, produced locally.

Iron Furnace, Mount Savage. Author photo.

The blowing engines at Mount Savage came from the West Point Foundry in New York in 1845, as had the ones for the furnace at Lonaconing. They were sized for furnaces making 400 tons of iron per week. Then engines were of the condensing type (recycling water), with a 56-inch diameter cylinder and a 10 foot stroke. They made 15 revolutions per minute, producing steam at 60 pounds per square inch and generating 80 horsepower. The associated boilers were 60 inches in diameter and 24 feet long. The grates spanned a total of 198 square feet. The blast cylinders were massive, being 126 inches in diameter with a 10 foot stroke. They operated at 15 revolutions per minute, and supplied air at 4-5 pounds per square inch pressure. One engine was used for the blast furnaces, and the other for the rolling mill. At the time, they were the largest cast cylinders in the world.

Early experiments with a coke-fueled furnace at Mount Savage in 1842 had produced acceptable iron at a cost of $16. per ton, when English iron was available for $15.84. A tariff bill, passed by Congress in 1846, removed protective duties on imported iron products. This benefited the English and Welsh manufacturers, at the expense of the fledgling American shops. This action was a direct cause of the failure of the Maryland & New York Iron & Coal Company, owner of the facility at Mount Savage. A poor showing for the facility that was, at one time, the largest manufacturer of iron in the United States. Mount Savage, at the time, represented one of the largest and most successful technology research and development facilities in the country, if not the hemisphere.

The 1846 Walker tariff was a United States Democratic Party-passed bill that reversed the high rates of tariffs imposed by the Whig-backed "Black Tariff" of 1842 under president John Tyler. A tariff (sometimes known as a customs duty) is a tax on imported or exported goods. The Tariff of 1842, or Black Tariff as it became known, was a protectionist tariff schedule adopted in the United States to reverse the effects of the Compromise Tariff of 1833.

The Walker tariff act was named after Robert J. Walker, who was formerly a Democratic Senator from Mississippi and served as Secretary of the Treasury under president James K. Polk. The tariff's reductions coincided with Britain's repeal of the Corn Laws earlier that year, leading to a decline in protection in both. The Corn Laws, in force between 1815 and 1846, were import tariffs ostensibly designed to protect British farmers and landowners, against competition from cheap foreign grain imports.

Its adoption was seen to increase commerce between the United States and

Britain. It was also predicted that a reduction in overall tariff rates would stimulate overall trade, and with it imports. The result, asserted Walker, would be a net increase in tax revenue despite a reduction in the rates. The Kane Letter by James K. Polk outlined his beliefs on tariffs, free trade, and protectionism during his 1844 campaign for President of the United States, and was widely circulated.

The Democratic-controlled Congress quickly acted on Walker's recommendations. The Walker Tariff bill produced the nation's first standardized tariff by categorizing goods into distinct schedules at identified ad-valorem rates rather than assigning individual taxes to imports on a case by case basis. The bill reduced rates across the board on most major import items save luxury goods such as tobacco and alcohol. An ad-valorem tax is a tax based on the assessed value of real estate or personal property.

Shortly after his election President Polk asserted that the reduction of the "Black Tariff" of 1842 would constitute the first of the "four great measures" that would define his administration. This proposal was intended to be the fulfillment of his campaign pledge in the Kane Letter on tariff policy that contributed to his victory in 1844 over Henry Clay. In 1846 Polk delivered his tariff proposal, designed by Walker, to Congress. The Tariff of 1842 placed a duty on pig iron of $9. per ton, and for manufactured iron, $25. per ton. The ad-valorum tax was set at 30%. Higher tariffs were good for the iron industry, but bad for the railroads.

The bill resulted in a moderate reduction in many tariff rates and was considered a success in that it stimulated trade and brought needed revenue into the U.S. Treasury, as well as improved relations with Britain that had soured over the Oregon boundary dispute. As Walker predicted, the new tariff stimulated revenue intake from $30 million annually under the Black Tariff in 1845 to almost $45 million annually by 1850. Exports to and imports from Britain rose rapidly in 1847 as both countries lowered their tariff barriers against each other.

The 1846 tariff rates initiated a fourteen year period of relative free trade by nineteenth century standards lasting until 1860. It was passed along with a series of financial reforms proposed by Walker.

The Walker Tariff remained in effect until the protectionism, which reduced rates further. Both were reversed in 1861 with the adoption of the Republican-backed Morrill Tariff.

Producers from other traditional protectionist constituencies such as iron and glass production, and sheep farmers, opposed the bill. When the Panic of 1857 struck later that year, protectionists, led by economist Henry C. Carey, blamed the downturn recession on the new Tariff schedule. Though economists today reject this explanation, Carey's arguments rejuvenated the protectionist movement and prompted renewed calls for a tariff increase. The Tariff of 1857's cuts lasted only three years. In 1861, the country changed course again under the heavily protectionist Morrill Tariff. But it was too late for Maryland & New York Iron & Coal.

Production

In the 1850's, the blast furnaces of Mount Savage had blazed around the clock, consuming massive amounts of coked coal, iron ore, scrap iron, and limestone. According to surviving records, in June of 1856, 356.5 tons of iron were produced. This required 747 tons of iron ore (at a 39 percent yield), 1.77 tons of coke per ton of iron produced, and 1.19 tons of limestone per ton. For this process, 536 tons of coal went to the blast engines. All of this raw material was dug by hand. The cost of production totaled $23.39 per ton, including anticipated repairs to the furnace, and wages. In 1844, number 2 furnace was in blast for 40 weeks, and produced 4,500 tons of iron. In 1846, number 1 furnace was in blast for 44 weeks, and produced over 4,500 tons. The integrated manufacturing center at Mount Savage, with its associated transportation infrastructure, represented the very cutting edge of the Industrial Revolution in America, and rivaled the best in the world. Economic issues, not technological ones, forced the shutting down of the blast furnaces in the late 1840's, but it re-opened during the Civil War.

From 1840-1860, profits in the iron business ranged from 40-60 percent, sometimes reaching 100 percent. Maryland was seventh in the nation in iron production in 1860, rising to fifth by 1870. The production of iron in Maryland declined sharply after that. The Maryland ore was never that good, and the discovery of rich veins in the West put the smaller, locally furnaces out of the iron business. Ruins of two of the furnaces are still visible in the town of Mount Savage, The Mount Savage blast furnaces had their own railroad branch, extending 1.3 miles from the main line. Bits of this roadbed can still be found.

The C & P Railroad shops employed 250 to 600 men, and another several hundred were busy at the fire brick works. The railroad operating crews were the best paid in the region. In 1880, Engineers got $3.50 per day, conductors $2.50, firemen $2.10, and brakemen $1.95. Boilermakers in the shops got $2.10 a day.

A skilled worker, in 1880, earned $2.00 for a day's work of 12 hours in the winter and 10 hours in the summer. He had to work six days a week with only Sundays and Christmas day off. An ordinary laborer earned $1.25 for a single day's work.

The brickyard was lucky to get one John Davis, a Welsh former rolling mill worker. He was of tremendous height and strength. His job was to straighten the crooked rails with a 100 pound mall. He was the only man who could lift and swing it. The mall is still at the brickyard and it is referred to as "The John Davis". Davis is Mount Savage's "John Henry."

The following page shows the production of iron at Furnace 1, and the raw materials used, on June 23, 1856. Meticulous records were kept by the iron master. A few have survived.

At this time, there was no way to measure the temperatures involved in the furnaces, except for the Iron Master's best guesses, based on his experience.

Wrought Iron

Wrought iron is made by reheating pig iron. One facility for doing this is called a puddling furnace, and this was the method used at Mount Savage.

Pig iron is then remelted and a strong current of air is directed over it while it is being stirred or agitated. This causes the dissolved impurities (such as silicon) to be thoroughly oxidized. The metal is then cast into molds or used in other processes. A puddling furnace, fired by coke or by the gas from coke production can produce wrought iron, which can be made into sheets or bars. The bars can then be rolled into rails.

Wrought iron is commercially pure iron. In contrast to steel, it has a very low carbon content. It is a fibrous material due to the slag inclusions (a normal constituent). This is also what gives it a "grain" resembling wood, which is visible when it is etched or bent to the point of failure. Wrought iron is tough, malleable, ductile, and strong in tension. Examples of items that are produced from wrought iron include: rivets, chains, railway couplings, water and steam pipes, raw material for manufacturing of steel, nuts, bolts, horseshoes, handrails, straps for timber roof trusses, boiler tubes, and ornamental ironwork. These are all pieces that can be used to make a locomotive.

Rolling the rails

The iron furnaces produced cast iron, an intermediate product. Early rails were cast, but these were prone to fracture. Pig iron is good for some things, but a better product was needed. This section examines the next step – a value-added process to turn pig iron into rails and motive power. A forge and rolling mill had been planned for the facility at Lonaconing, but these were never built.

In 1842, the American Railroad Journal had said in an editorial that there was no firm in the United States capable of manufacturing heavy-edged rail. Many facilities had tried and failed to produce an acceptable product. The market was apparent and the Mount Savage rolling mill was built in 1843 by the Maryland & New York Iron & Coal Company. The rolling mill site had 3 trains of rollers driven by steam engines, 17 puddling furnaces, 6 reheating furnaces, and 3 special facilities for sheet iron production. The furnaces were of the Siemens type, using coal gas produced on site as a byproduct of coke production as fuel.

A medal for the process was awarded in October 1844 by the Franklin Institute of Philadelphia. The medal was at one time a part of the collection of the Museum of Ince Blundell in Lancashire, England. In 1844, there was just over 4,000 miles of railroad in America.

The Mount Savage open hearth furnaces were not making true steel – something between malleable iron and mild steel. The difference between cast iron, wrought iron, and steel is a percentage point or two of carbon. The rival Bessemer steel process was patented in England in 1856. It was the first inexpensive way to produce steel from pig iron. Air is blown through the molten iron, removing impurities such as silicon, manganese, phosphorus, and excess carbon. The Bessemer process was superseded by the open hearth furnace by the end of the century.

William and Frederick Siemens built their first experimental furnace in 1858, and got a patent in 1861. By 1868, they finally demonstrated the production of steel from pig iron.

The Siemens puddling furnace was a rectangular, covered design that passed burning gas over the top of the charge of pig iron. The early models had a 4-5 ton capacity. The concept was to purify the pig iron by oxidation of the carbon, and to remove impurities. The contact with the products of combustion can be controlled by controlling the draft, and it can be used to add or remove material from the melt. It can produce mild steel from pig iron by lowering the carbon content.

Puddling furnaces were making malleable iron out of pig iron in England as early as 1784. In these, the molten metal with a layer of floating slag is stirred for 5-10 minutes of clearing. This causes the oxidation of contaminates in the iron, mainly silicon, manganese, and phosphorous. The temperature in the furnace was raised to the boiling point of iron, 4,982 degrees F. The puddler's job was a particularly hot and dangerous one. He stirred the molten iron with a long rod, to bring the slag and impurities to the surface, where they could be skimmed off. Still, it was probably a better job than in the nearby mines. His skill was also critical. There was no way to accurately measure the temperature of the melt, except by its color. The ironmaster's experience and calibrated eyeballs were the key to success or failure of the process.

After clearing, the temperature in the furnace was raised. After another 10 minutes of stirring, carbon monoxide gas would begin to escape the melt as bubbles, which catch fire as they burst. This phenomenon was called puddler's candles. When they disappeared, more stirring was called for. This got increasingly difficult, as globules of decarburized iron formed. These 150 pound semi plastic balls were grasped by tongs and removed from the furnace. These were then hammered or squeezed by rollers. The resulting rough bars were cut into short pieces, and went into a reheating furnace. Here, they were taken to white heat – a self-welding temperature. They would be worked and reheated numerous times. Each reworking with a hammer or rollers removed more slag.

For the rolling operation, the worked iron was again reheated, and shoved by steam power through the rolls multiple times to get the correct shape. The Mount Savage rail of 1844 was a U shaped design. Rolled rail would have a natural curvature as it cooled, and would be bent until straight. The rail would then be weighed for quality purposes, and stamped. Later, heat-treating of the working face would be used. Plate was also rolled in a similar process to rails. It would be sheared to size, and holes for rivets would be punched.

The production of coal gas is a byproduct of coke production, but in carefully designed reactors, a better grade of gas is produced. Typically, a producer reaction vessel is used, being constructed of cylindrical steel like a boiler, and lined with firebrick. It was partially filled with coal or coke. A blast of air, and sometimes steam, was introduced from the bottom. These reactors consume several tons of coal per hour.

Quoting Pangborne in his 1894 book, "The first rail rolled in America is turned out in 1844 by a mill at Mount Savage, Allegheny County, Maryland, and the Franklin Institute of Philadelphia, in honor of this event, has struck a silver metal. The rail is of the U form, and laid on a longitudinal sill, to which it is fashioned by an iron wedge, keyed under the sill, thus dispensing with outside fastenings. The rail weighs forty-two pounds (to the yard), has been introduced on the line between Mount Savage and Cumberland, and also on the Baltimore and Ohio road. The mill has an order for rails weighing fifty-two pounds to the yard from the road from Fall River to Boston."

Z-rail

Benjamin Latrobe, of the B&O Railway, patented an interesting product called Z-rail. This was mentioned in the American Railroad Journal, No. 6, Vol. VI, 1837. Some was made in Mount Savage for the Eckhart Rail Road, then under construction. The Railroad Journal mentioned, "This form of rail has not yet been fairly tested, but we have much confidence in its success." It evidently didn't work out.

A successful rolling mill had been placed in operation in England as early as 1783, and cold rolling began in Pittsburgh in 1860.

In addition to rolled product such as rail, the production facility could produce wrought iron sheet. Rolled into cylinders and riveted, this made the production of locomotive boilers possible. In addition, wrought iron was used in the production of chain, engine bolts, stay bolts, pipe and threaded parts, and drawbars.

Old Rail Mill

Erected at Mt. Savage, Allegany county< Maryland in 1843. In this Mill was rolled, in 1844, the First Solid Track Rail made in the United States of America.

Courtesy, Howard Redford Aldritch

The first successful output of the Mount Savage mill was in 1844, and marked the end of the U. S. dependence on imported products. Interestingly, in the same year, iron ore was discovered around Lake Superior. This would lead to the demise of Mount Savage in particular and Maryland in general, as world-class iron producers.

The 43 pounds to the yard rail was used for the home road and sold to the B&O railroad, which up to that time had been dependent on imported British rail. One thousand tons of rail, at $59 per ton, went to a railroad at Fall River, Massachusetts. Most interestingly, the Mount Savage facility used its own product to build the Mount Savage Rail Road, down the Jennings Run and through the Narrows, to connect with the B&O at Cumberland. An additional customer included the Hampshire Coal & Iron Company for their tram road near Piedmont, WV. The Utica & Schenectady and the Hudson River Railroad in New York, and the Erie and the Reading in Pennsylvania were also customers. The Utica and

Schenectady Railroad ordered 1,000 tons of rail. There was a display of Mount Savage rail at the Mechanics Fair in Baltimore in November 1850. E. Pratt & Brother were the agents.

During the Civil War, the facilities at Mount Savage went back to work. Some 333 tons of railroad iron from Mount Savage were captured by US forces in 1864 in Louisiana. The government reimbursed the company for the cost. The rail had originally been sold to the El Paso Pacific Company in Texas, but never used. Some 53 tons were lost to Confederate forces. Railroad iron was used as makeshift armor plate for ships.

John W. Brown of Mount Savage was granted a patent for a T-iron rail rolling mill in April 1856. It describes a five-step process, which not only forms bar iron into rail, but controlled the displacement and density of the finished shape, and the hardness of the wearing surfaces.

In the 1850's, the Mount Savage facilities employed 900. Jobs at Mount Savage attracted both skilled and unskilled immigrants from Ireland, England, and Wales. In foundries and machine shops wages were relatively stable from the early 1880's to the 1910's. A machinist or boilermaker would make about $2.50 per 10-hour day, 6 days a week, 300 days per year. There was no vacation, no sick leave, no holidays except for Christmas day. The employees were paid in script, exchangeable at the company store, until the 1880's

The firm of Manning & Lee, Charles & Towsend Streets, Baltimore, were Agents for the Mount Savage works. In 1845 and 1846, they brokered the sale of "T" rail to a Boston Purchaser. This is probably the delivery for the Fall River, Massachusetts, railroad.

Industrial safety was a concept that developed slowly and the iron shops of the 19th century were dangerous places to work. As part of the social contract men injured at work usually had guaranteed lifetime employment -- if they survived.

Where many hundreds of men labored in literally hellish conditions, a stranglehold on the rail industry by British industry was broken, and the tools for enabling the industrial expansion of the United States were produced.

Steel

Steel is made with just the right amount of carbon added to the reheated iron. The trick is getting the right temperature and the right amount of carbon, without other contaminants. Small batches of steel were made in the Middle ages, possibly back as far as 500 BC, in India.

Being able to make steel in large quantities was the goal. There are several methods. Steel differs from iron in the carbon content, which must be carefully controlled. Steel is an alloy of iron with a carbon content between 0.2% and 2.14%. The Bessemer converter in the mid-19th century lead to the mass production of steel. The Mount Savage Works were on the verge of making steel, before Bessemer got his process working in 1858. The first Bessemer steel mill in the United States was established in 1855 in Wyandotte, Michigan, about 14 miles south of Detroit.

A later alternative to the Bessemer process is the open hearth furnace process, using the Siemens regenerative furnace that Mount Savage is known to have had. A temperature of 1600 degrees F must be sustained. A Siemens furnace was first used for steel production in France is 1865. This became the preferred method of making steel.

What if? What if Mount Savage had produced the first economical steel in the United States? Would Pittsburgh still have developed as the steel-making capitol of the United States? Would Mount Savage be cleaning up after a Century of Industrial pollution?

Mount Savage Railroad

The Maryland & New York Iron & Coal Co. was charted in 1838, and built several blast furnaces at Mount Savage, Allegany County, Maryland. These furnaces were modeled on the Georges Creek Coal & Iron Company's Lonaconing Furnace. Besides the blast furnaces, facilities were built to work the cast iron, most notably a rolling mill, where the first American made iron rail was manufactured in 1844. Five hundred tons of rail were produced for the company's railroad, following the path of Jennings Run to Will's Creek, and through the Narrows to the B&O railhead at Cumberland. Bridges were originally built of timber, but were later replaced by iron due to excessive maintenance costs.

In February 1844, records indicate that the B&O railroad supplied engines and cars to the Maryland & New York Iron & Coal Company. The 10 mile long Mount Savage Rail Road was completed to Cumberland in 1845, the same year Florida was admitted as a state. The bridge over Will's Creek west of the Narrows towards Mount Savage is dated 1842. The B&O did not send their best equipment into mine service. Engines of the "second class" were used. This classification was based on weight and performance, not quality. To put it in better terms,

"It must be remarked that the duty of the 2nd class engines appears so much less than that of the other classes, not from inferior efficiency, but from circumstances which have given the two engines of this class less to do than they could have accomplished. This is particularly to be said of the engine of this class which has done the work of the Mount Savage Road; this engine being, in fact, one of the best in the service." (21st. Annual Report of the B&O, Oct. 1847, p. 43)

In a week in December 1852, the Mount Savage Rail Road moved more than 5,500 tons of coal to Cumberland. The first commercial contract signed by the Baltimore & Ohio Railroad to move coal was signed in February 1844 for the Mount Savage product.

April 1, 1845, marked the date of an historic agreement between the B&O Railroad and the Maryland & New York Iron & Coal Co. It stipulated a charge of 1 1/3 cents per ton-mile to transport coal from Cumberland to Baltimore, provided the company shipped at least 175 tons/day for at least 300 days of the year. Coal was still viewed as a speculative commodity by the B&O. Wood or charcoal were the fuel of choice for industry, and for home heating. Connection was made with the B&O in Cumberland in 1846. Also in that year, the B&O contracted for 15 miles of Mount Savage rail, nearly 675 tons of the 51 pounds per yard product then produced. The rail was used to upgrade the line between Harper's Ferry and Baltimore.

Before this purchase, the B&O was relying totally on imported British rail.

Ross Winans of Baltimore supplied engines and rolling stock to the Mount Savage Rail Road. It is not known if this is a complete list of Mount Savage motive power. All of the listed engines are of the 0-8-0 'Camel' type. Four of the engines went to the C & P Railroad.

Mount Savage Railroad Motive Power Roster

Builder	type	date	Name	company	
Winans	0-8-0	1848	Mount Savage	MSRR	rblt 1868, scrp 1891
Winans	0-8-0	1850	New York	Mount S. Iron	unknown
Winans	0-8-0	1852	Highlander	Mount S. Iron	rblt 1868, scrp 1891
Winans	0-8-0	1852	Frostburg	Mount S. Iron	rblt 1866, 75. scrp 1896
Winans	0-8-0	1853	Galloway Lynn	Mount S. Iron	rblt 1874, scrp 1896

Before 1851, general railroad practice was to name engines rather than number them. Locomotives were generally named after geographical references, or persons of significance.

In 1845, the railroad inaugurated passenger service from Mount Savage, with connections to the B&O in Cumberland. Three trains per day were provided and operated by the B&O. At that time, the trip from Baltimore took 8 1/2 hours. William Cullen Bryant wrote of his trip in the Saturday Evening Post, providing a fascinating glimpse into the rigors of the early travels. He writes, "At Cumberland, you leave the B&O railroad, and enter a single passenger car at the end of a long row of empty coal wagons, which are slowly dragged up a rocky pass beside a shallow stream into the coal regions of the Alleghenies."

The Maryland & New York Iron & Coal Co. failed in 1848, and the assets were sold to the Lulworth Iron Company. The Mount Savage Rail Road was transferred to Lulworth on January 14, 1848. That same year, Lulworth itself reorganized to become the Mount Savage Iron Company.

Winans delivered ten coal cars to the company on August 1, 1850, followed by five more in September. He had worked on 6 of the 4-wheel cars previously, in August of 1845. The Mount Savage Iron Company completed the Canal Wharf in Cumberland in 1850.

The Mount Savage Iron Company had extended their rail line northward 3 miles from Mount Savage to Borden Yard in 1851. The main line was double tracked from Cumberland to Mount Savage Junction in 1872, and a third track was added in 1902. Travel west on Route 40 through the Narrows from Cumberland, and the old Western Maryland line, now used by the Western Maryland Scenic RR, will be on your left. The CSX main line across Will's Creek on the right used to be Mount Savage Rail Road, then C & P. The B&O had trackage rights on these lines. The Pennsylvania Railroad came from Bedford near the west end of the Narrows on a bridge, since removed. This would become the Western Maryland State Line Branch, but the PRR once serviced the passenger trade from the original WM station. At the West end of the Narrows, C & P's Eckhart branch split off, and passed under the Western Maryland iron truss bridge, circa 1907.

Mount Savage, operating center of the railroad, is located at milepost 9.4. At Mount Savage were the C & P shops, roundhouse, passenger station, a wye, and the Mount Savage Refractories brickworks. The roundhouse suffered a fire in 1907, and the depot, which resembled the ones in Frostburg and Lonaconing, was dismantled in 1955. The shop's turntable was not powered, except by strong arms and backs. There were car shops, a freight house and yards, an oil house, machine and pattern shops, and the office building. The Mount Savage town park is located roughly on the site of the roundhouse. The Mount Savage coal ramp was used by the Pioneer Coal Corp., the Cumberland Parker Seam Coal Corp., and the Victory Coal Mines. Five additional companies loaded coal by conveyor (circa 1947). Earlier, the shop's coal tipple serviced local needs. When in full use, the Mount Savage blast furnaces required 150 tons of coal per day.

Beyond Mount Savage is Morantown, Allegany (or, Zihlman), and Borden Yard. One mile beyond that is Switch Number Nine. Allegany had a depot, water tank, and freight shed. Borden Yard also had a station, and was the location for a major brickyard of the Big Savage Refractories.

The C & P Railroad acquired the Mount Savage Rail Road operation in January 1854. This acquisition included the motive power, rolling stock, 14.9 miles of track from Borden to Cumberland, and the Canal Wharf. Thus, the Mount Savage Rail Road disappeared as a separate corporate entity, but became the basis for the next generation of short line railroads of Allegany County, Maryland, and a direct ancestor of present-day CSX Transportation. The critical path through the Cumberland Narrows was built, as we have seen, by the Mount Savage Rail Road. It is still in daily use, as 12,000 horsepower diesel multiple units routinely haul commodities up the Sand Patch grade and through the tunnel, to Pittsburgh, Chicago, and the west.

Cumberland & Pennsylvania Railroad

The major C & P facilities were in Mount Savage, the heart of the railroad. These included a sand house, an engine house (1888), carpenter shop and supply house, an electrical supply building, a car inspector's office, hose house, and a lumber store. The office building, built in 1902 of locally produced enameled brick, still stands. The brick roundhouse, circa 1907, had a sixty foot deck 'Armstrong' turntable. There was an associated oil house and supply shed. The stone machine shop and car shop from 1866 still survive. The car shop was extended with a wooden structure in 1906.

The blacksmith shop dated from 1900. There was a paint shop, and paint supply house. Next to the two hundred foot long passenger car shed was the tin and upholstery shop. There was also an ash pit and trestle. The power house at the Mount Savage works, circa 1908, included dual 50 kilowatt, 250 volt generators. In addition, there were two large air compressors for the shop tools, a fire pump, a boiler feed pump, and a converted locomotive boiler, 53" in diameter, and 21 feet long.

The Mount Savage Roundhouse suffered a fire in 1907. The roundhouse suffered extensive damage, but was rebuilt. There was a fire in the car shop in 1939.

The railroad is more than rail and locomotives. The Mount Savage Shops produced other equipment as well.

C & P engines spent as much time running backwards as forwards, so the tenders were typically equipped with lamps and pilots. Two major types of tenders are in evidence. The earlier units have a large box-type oil lamp, and a 'cowcatcher' on the rear. The newer types had an electric lamp, and the cowcatcher is gone. The coal capacity was generally 7-9 tons, and the water capacity was 5-6000 gallons. The engines were always within reach of a source of coal, were usually close to a water supply, and were never far from the home shops.

Not surprisingly, most of the C & P Freight equipment consisted of coal hoppers. The 1923 valuation lists 2,500 freight cars, of which 2,489 are designed to haul coal.

Several types of gondola cars are listed on the roster. Thirteen cars in the number series 616-630 were built by the C & P during 1895-1898, and are of wooden construction. These are listed as 50,000 pounds capacity, and thirty-five feet long.

Maintenance of way (MoW) is necessary for any railroad to maintain the infrastructure, and to respond to accidents and equipment breakdowns. C & P had its own maintenance of way equipment, and, like most railroads, used older equipment pulled from revenue service, and custom rolling stock. The wreck crane, 'Big Hook' number 105 was kept in Mount Savage. This was a 1902 steam unit with steel body and underframe. Besides its use in wreck clean-up, the C & P Big Hook was rented out to assist in heavy lift tasks. After the new windlass for the deep mine at Ocean arrived by flat car, the C & P hook was used to set it into place. The crane was also used to position the road bridge over Wills Creek at Locust Grove, at the West end of the Narrows.

A smaller crane, numbered 102, was constructed by the C & P in 1901, and consisted of a 40,000 capacity wood underframe car equipped with a 5 ton manual crane. It was kept in Franklin. No pictures are known to exist. There were two riggers' flats, No. 103 and 104, 40,000 capacity, built by the C & P in 1904 of wooden construction. These held auxiliary cables and blocks, and sometimes an old tender water tank for the crane. There was also an electric service car, built by C & P in 1889, and numbered 106, based on a 50 foot wooden car. The unique C & P tool car, number 109, was based on a 60,000 lb. capacity wooden car with composite underframe. C & P special car No. 605 was used as a tunnel icebreaker.

There are rumors of a custom snow plow from the Mount Savage shops, but no photos or descriptions have been found. Equipment lists filed with the ICC also show two circa-1898 converted hoppers, numbered 107 & 108, 50,000 lb. capacity, and steel construction. These were typically used for ballast. The ICC roster also lists a cement car, no. 31. This was built by the C & P in 1894, and was a 37 foot wooden unit, with four-wheel arch-bar trucks. Track cars were based at Mount Savage yards and Carlos Junction. One hand car and four push cars were kept at Eckhart.

Ten hand cars, and fifteen push cars were rostered in 1917. Also included with the maintenance equipment were a concrete mixer, two narrow-gauge concrete cars, and a Climax stone crusher.

C & P had a valuable asset in the car and locomotive shops in Mount Savage. With the facilities and experience for locomotive and rolling stock rebuilding and repair, C & P was to a great degree self-sufficient. This allowed the recovery of assets that might otherwise have to be abandoned. Even competitors such as the Georges Creek and Cumberland used the shops when necessary, such as the GC&C accident in January 1883 that sent two of its ten locomotives and fifty-one cars off the trestle at Vale Summit. The locomotives were recovered and sent to Mount Savage for repair.

The C & P had two business cars, the first being constructed in the home shops in 1899. Number 15 was 37 feet in length, with a wooden body and underframe, and composite, 4 wheel trucks. Number 101 was purchased from the Pullman Company, and was 75' 6" feet long, with a steel over wooden sheathing body. The frame was metal, and six-wheel trucks with 36" wheels were used. This car was valued at more than $26,000. It was equipped with Westinghouse air brakes, and hot water/steam heat. It had a 60 volt electrical system, using batteries and axle-driven generators. The car had a water supply system, and sported a kitchen, dining room, state rooms, staff quarters, and an observation room. The car was painted Pullman Green, and the interior featured Mexican and Cuban Mahogany. Unfortunately, it would not clear the Frostburg tunnel.

Most of the C & P cabooses were four-wheel units, built at Mount Savage. The 1922 ICC Valuation lists nineteen of these in service, valued at $640. each. They were of wooden construction. The design followed those of the Pennsylvania Railroad and the Western Maryland. Cabooses No. 33 and 34 were 37-foot, wooden bodied, eight wheel units. These were probably older converted passenger coaches. No pictures are known to exist of these units. They were built by the C & P in 1894 and 1898. Early C & P four wheel cabs had cupolas. Following Western Maryland practice, later units did not. No C & P cabooses were involved in the Western Maryland merger in 1953. Four C & P cabs were rostered as Western Maryland Maintenance of Way equipment, including two converted to weed burners. All were scrapped before 1953.

Ballast for the line was crushed stone from Kreigbaum, and cinders and granulated slag from Mount Savage.

Locomotive Mount Savage

C & P Railroad Locomotive Number 1 was named *Mount Savage*. This engine was built by Ross Winans of Baltimore. It was a Camel design, wheel arrangement 0-8-0. It had been purchased in 1848 by the Mount Savage Rail Road. It weighed 50,400 pounds, and had 17" x 22" cylinders. Rebuilt at the Mount Savage shops in 1868, it was scrapped in 1891. No pictures of this particular unit are known to exist.

Steamship Mount Savage

The *Mount Savage*, a 452-ton (burden) screw steamship, was built in 1853 at Philadelphia, Pennsylvania. The vessel was sold at auction to Mr. A. C. Hill in 1854 for $10,000. She had cost $40,000 to build. She was renamed *Memphis* in 1857. Chartered by the Navy in September 1858, she served as *USS Memphis* during the Paraguay expedition of late 1858 and early 1859. The steamer was purchased by the Navy in May 1859 and renamed *Mystic* a few weeks later. In June and July 1860, while operating off Africa, she captured two slave ships.

During the first part of the Civil War *Mystic* served in the blockade of the Confederacy's Atlantic Coast. She assisted in the capture or destruction of four blockade runners off North Carolina in June-September 1862, among them the steamers *Emma* and *Sunbeam*. While in the process of taking the latter, on 28 September, she was damaged in collision with USS *State of Georgia*. *Mystic* was employed in the Chesapeake Bay region from late 1862 until the war's end. In May 1863 she supported the Army during an expedition up the York River and in September of that year seized a sailing vessel off Yorktown. USS *Mystic* was sold to private owners in June 1865. Renamed *General Custer*, she disappeared from merchant vessel registers in 1868. Her disposition is unknown.

Opening of the Mount Savage Railroad Extended

Miners' Journal, Cumberland, MD, September 24, 1852

"Monday, the 20th inst. being the day appointed by the Managers of the Mount Savage Iron Company for the opening of the Extension Railroad, connecting the present termination of the Mount Savage Railroad with the mines of the various Coal companies in the neighborhood of Frostburg, everything was most hospitably arranged for the reception and accommodation of the various guests invited from the vicinity; and a large and happy party of ladies and gentlemen assembled in Cumberland, and entered the cars prepared for them.

It may not be amiss here to give our readers a short description of the Mount Savage Railroad, now the principal feeder of the Baltimore and Ohio Railroad and the Chesapeake and Ohio Canal, with the renowned semi-bituminous coal of this region. It leaves Cumberland, where it unites with the Canal and Baltimore Railroad by separate branches, passing within an easy grade through the sublime scenery of the Narrows of Wills' Mountain, which is crested with a granular white stone of pure silex, destined, at no distant day, to add to the wealth and population of Cumberland in the manufacture of every kind of glass, for which it is admirably calculated.

At the western entrance to the gorge, two miles from Cumberland, the Railroad is intersected by one leading up the valley of Braddock's Run, to the Eckhart, Washington, Pompey Smash, and other valuable Mines. This Road, built by the Maryland Mining Company, as we; as the greater part of the coal property just mentioned, is now owned by the Cumberland Coal and Iron Company.

From this point, the Mount Savage Road proceeds two miles up Wills Creek, crossing that stream by a handsome timber bridge of two arches, each being eighty-five feet span. The gradual decay of timber requiring constant repair, the Company have it in contemplation to erect, in its place, an elegant and substantial iron bridge.

On reaching the mouth of Jennon's (sic) Run, the heavy grade of one hundred feet to the mile commences; and in three miles, at the elevation of nine hundred and eighty feet above tide water, this iron artery first enters the Coal field, to draw from thence the life of a thousand gigantic ocean steamers, ploughing their unvarying, rapid course, in spite of wind and tide, to every known quarter of the world.

The first mines reached are those of the Parker Vein Coal Company, at Barrellville, on the north fork of Jennon's (sic) Run, under the management of M. P. O'Hern, Esq. A Railroad, three quarters of a mile in length, and spanning Jennon's (sic) Run by an elegant Viaduct forty feet in height and three hundred in length, connects them with the main Road, while their little mining village, with its clean and pretty cottages, presents a picturesque appearance. The Parker Vein was the first discovered and earliest mined in the Coal field, and was boated down the Potomac for use in the Government Arsenal at Harper's Ferry, long before the iron age of Railroads had commenced.

Two miles further brings us to Mount Savage-the largest Iron Works in the United States.

Three immense furnaces, which, when in blast, can produce fifteen thousand tons of pig iron a year-a large and most complete foundry and machine shop-a noble rolling mill, capable of turning out ten thousand tons of rails or bar iron a year, or even more-a manufactory of the best fire brick in the United States-dwellings for twelve hundred operatives and their families, estimated at a population of fully five thousand-all this bursts on the view as we arrive at the beautifully situated village of Mount Savage. But now, from the want of protection to the Iron interests of this country, and the low rate of wages abroad, the dwellings are deserted, and the silent furnaces and rolling mill serve only to show how much injury has been effected by this most unwise policy. It is hoped, however, that the recent rise in iron will again put these works in operation, which would prove an immense boon to the working population in the neighborhood.

Here commences the continuation of the Railroad.

It has been extended to five miles in length, so as to overcome the elevation by a grade not exceeding one hundred and five feet to the mile. Some of the embankments are very heavy; the whole road is of a most substantial character, thoroughly ballasted, and laid with Winslow's Patent heavy compound Rail of 65 pounds to the yard. It is needless to add that the Rails were made at Mount Savage. The road reflects the highest credit on C.F. Fogg, the Engineer, to whom was entrusted the entire management.

The first mines passed are those of the Alleghany Company, E. K. Huntley, Esq., Superintendent; and one mile further, at the foot of Frostburg Hill, are the Depots of the Frostburg Coal Company, under the superintendence of D. C. Bruce, Esq., and the Borden Company, under that of A. C. Greene, Esq. Besides the Coal companies already at work, are many others-the Wilthers, the New York, Cumberland Coal and Iron Company, &c.-all possessing large and valuable coal mines, which, at a future day, will tend to swell the amount of transportation on this Railroad. The Cumberland Coal and Iron Company have a large and valuable portion of their coal property coming out on this valley at this point, connecting with their larger deposits on George's Creek; and it is contemplated by the Mount Savage Company, at no distant day, to construct a locomotive tunnel through the big vein under Frostburg Hill, to the valley of George's Creek- a comparatively short distance; which would open a market to an immense area of coal field, so as to bring, by this route, the mineral deposits extending to Lonaconing, which would otherwise have no other outlet than the Railroad down George's Creek, through Westernport, and then over twenty-seven miles of the Baltimore and Ohio Railroad to Cumberland.

The present weekly transportation of Coal, down the Mount Savage Railroad, is nearly six thousand tons; and the greatly increased facilities afforded by the opening of this new line, will, in the course of the next few weeks, increase it to nearly ten thousand.

On the brow of the hill, overlooking the Railroad and depots, is the elegant Italian villa now building for A. C. Greene, Esq., and to this the company adjourned, to partake of a collation and champagne, generously provided for them by the Mount Savage Company.

The company then reentered the cars, and rapidly descending the grade, returned to Cumberland, highly gratified with their excursion."

Locomotive Shops

This section discusses the Mount Savage Locomotive shops.

Locomotive manufacturing

After the civil war, Master Mechanic James Millholland came down from Pennsylvania to set up the C&P shops. A period of locomotive rebuildings and experimentation followed, culminating in the production of engines at Mount Savage.

The C & P locomotive shops were established in Mount Savage in 1866, under the direction of James A. Millholland. The original locomotive shop was constructed of stone and was 90 feet x 250 feet in size with a 33 foot high roof. An adjoining car shop, built at about the same time, was also of stone and was later extended with a wooden structure. These buildings still stand and are in use in Mount Savage.

James Millholland, Senior, was 54 years old when he and his family came to Mount Savage from Reading, Pennsylvania. Millholland was an "advocate of plain engines and simplicity." He had extensive experience in keeping Winans camel engines running from his earlier work with the Baltimore & Susquehanna Railroad, and he was credited with many important locomotive innovations. He came in 1866 as the President of Consolidation Coal, and of the C & P Railroad. He resigned in 1869, retiring to his estate on the Valley Road in Cumberland. He was credited with developing the first anthracite burning locomotive, and was Superintendent of Motive Power for the line for many years. He is also credited with constructing the first iron deck girder bridge in the United States for the Baltimore & Susquehanna Railway near Bolton in 1846-47. He was responsible for so many improvements to the basic Winans camel engine, the class was referred to as "Millholland Camel's". He is also credited with designing a 12-wheeled camel engine, built in the

Pennsylvania & Reading shops in 1863.

His son, James A. was 24 years old when the family moved to Mount Savage. He had been born in Reading, in 1842, and had apprenticed in the railroad shops. He also joined the C & P, becoming Master Mechanic, and was vice-president by the time his father retired. He was also listed as the Second Vice President of the Georges Creek Coal & Iron Company in 1869. He left the C & P in 1879 to join the new Georges Creek & Cumberland Railroad. The younger Millholland had been tasked with building the C & P shops, to maintain the mixed fleet of motive power. He had the right experience for the job.

Millholland bought quality machine tools, which were still in use 40 years later as evidenced by the 1917 ICC valuation. He equipped the shops with metal working machinery from Bement & Dougherty, probably a predecessor of Wm. B. Bement & Son of Philadelphia.

Initially, the work supervised by Millholland at the Mount Savage Shops was limited to repairing and rebuilding the Winans Camels and other early C & P locomotives. The shop force gained valuable hands-on experience during the first twenty years; at least 15 of the C & P's Camel locomotives were rebuilt at Mount Savage (some twice). Typical of the rebuilds was the engine *Highlander*, a Winans Camel inherited from the Mount Savage Rail Road. This was a modernization project in which, among other things, the cab was relocated from on top of the boiler to a rear position. The C & P shops also provided repair services to its rivals in the Georges Creek coal region.

By the 1880's, the shops that Millholland had set up apparently had built quite an extensive operation, able to offer custom built locomotives for sale in addition to meeting the requirements of the parent line, the C & P. The period beginning in 1883 was an exciting one for heavy manufacturing in Mount Savage. A locomotive catalog listing five types of engines for sale and their specifications was issued for the Works by their agent, Thomas B. Inness & Co. of Broadway, New York. That Catalog has survived, and reprinted in the Author's *C&P Revisited* book.

Evidence was that the catalog was successful, and numerous sales to other roads resulted. This helped finance production for the home road, spurred development, and helped employment. Narrow gauge engines proved so popular a product that the Mount Savage works installed a third rail up the main line from Mount Savage for customer acceptance testing of narrow-gauge equipment.

The first recorded engine 'build' was a 0-10-0 unit in 1868. This could have been a modification to a Winans Camel. Engine production was active between 1885 and 1917. Engines were produced for other roads as well as the C & P. The production figures for 1882 list 19 passenger and freight engines outshopped, with 16 more in 1883.

A new 4-stall roundhouse was built in Mount Savage in 1898. It was reported then that the locomotive shops were working 5 hours of overtime every evening.

All of the rotating power machinery in the Shops were driven by leather belts from overhead master shafts. These, in turn, were powered by a stationery steam engine in the adjacent power house. A similar facility may be seen today preserved at the East Broad Top Railroad, in Pennsylvania.

One particularly good customer was T. H. Paul & Sons of Frostburg. A former C & P master mechanic himself (1854-1855), Paul established shops in Frostburg and Cumberland. He built mine engines and smaller narrow gauge locomotives at his shops, but contracted with Mount Savage for his larger orders. His Frostburg works were located near the existing C & P Passenger station on Depot Hill.

Locomotive manufacturing during this period was hard, heavy, dangerous work. It proceeded according to numerous 'rules of thumb' developed by the master mechanics over the years. Innovations were introduced slowly. There were continuous efforts to reduce costs, and increase performance. Weight reduction was not desirable, as weight-on-drivers contributed directly to tractive effort. Locomotive frames were usually riveted, built-up construction, of wrought iron and later steel.

According to White, experience at the Norris locomotive works showed that a team of 14 men could build a locomotive in 15 days. This was assuming the parts were on hand. A locomotive is a carefully integrated collection of a large number of specialty parts. The typical boiler was constructed of 5/16" wrought iron, starting as plate, and rolled to shape. The lap joints were single riveted. There is a long way between watertight and steam tight. Later, double riveting, and reinforced butt joints were used. Welding was not yet a developed technology, particularly for a pressure vessel. Boiler tubes were typically iron tubing of 2 inches diameter. They were lap welded, and reportedly difficult to flange.

The cylinders were usually cast in halves, assembled, and bored to size. This represented the most complex and expensive operation of the whole locomotive assembly. In 1856, it was common for the boring operation to consume 2 days. The pistons were cast iron, with fitted brass piston rings.

Boilers were covered, or lagged, to reduce hear loss, and increase efficiency. Wood slats were used originally. After 1900, asbestos was a favored lagging material. It was common for the slabs of the mineral to be machined to fit. This produced large clouds of asbestos dust that is now known to be a major carcinogen, a significant cause of lung cancer. The use of dust masks, hearing protection, and safety glasses was unknown at the time. The boiler shops operated in a haze of asbestos dust.

Millholland favored Giffard's water injectors, based on the favorable experience with them on the Reading line. He was also an early advocate of feedwater heaters, using them as early as 1855. His designs have them on the right side, under the engine running board. They are about 10 feet long, and 8" in diameter. These are a visible clue to engines produced in Mount Savage. Millholland is also responsible for the development of the poppet throttle, originally retrofitted on Camel engines in Pennsylvania.

The driving wheels were typically cast iron, and axles were usually 6" diameter wrought iron. Driving wheels were fitted with replaceable tires. On the basis of his previous experience, Millholland favored cast iron tires, shrunk onto the wheels. His father had experimented with steel tires around 1851-52, and they became standard later. Some early accidents on the C & P involved wheel failures. In 1872, Engine No. 11 broke a wheel below Frostburg, requiring the assistance of the work train, and delaying the pay car, according to the Frostburg Mining Journal.

Connecting rods were cast, and bearings were brass and/or Babbitt metal. The early lubricants were all animal fat based, and only suitable for low temperature applications. Later, petroleum based lubricants provided much better performance.

Engine safety appliances were sparse. The Bourdon Gauge for pressure readings was patented in 1849. A rival gauge was developed in 1857 by Wooten. The McKaig Company of Cumberland were producing steam pressure gauges. Glass sight gauges for boiler water level were not popular until the 1890's. Head lights were originally oil lamps. These units were box-shaped, and had a 18-22" parabolic reflector. They could cast a 1000' beam, sufficient for low-speed operation. An important improvement was introduced with the advent of lamps powered by carbide. Similar to the lamps used by miners, these lamps used the reaction of water and the mineral calcium carbide to produce acetylene gas, which burned with a bright light. Later, electric lamps and generators were fitted. C & P tenders were also fitted with lamps on the rear, since the engines frequently operated in reverse on the various coal branches where they could not easily be turned.

Shops Crew

Engine Rebuildings at Mount Savage Shops

The following table shows the documented engine rebuildings at Mount Savage. Millholland had extensive experience with rebuilding and upgrading the Winans design. This list is derived from Hicks. It covers the major engine rebuildings for the home road, not routine maintenance or quick operational repairs. H-B refers to the manufacturer Hayward-Bartlett.

C & P Engine Rebuildings at Mount Savage

No.	Original	Date	Notes
3	Winans	1866-75	weight increase
10	Winans	1866-75	
12	Winans	1866-75	
13	Winans	1866-75	
1	Winans	12/1868	weight increase
2	Winans	6/1868	weight increase
22	Winans	1870	due to boiler explosion

No.	Builder	Date	Notes
23	Winans	1870	
4	Winans	1874	
31	Baldwin	1879	
5(32)	Baldwin	1885	
14	H-B	1887	
25	Baldwin	1888	
15	H-B	1888	
16	H-B	1889	
17	Norris	1898	
25	Baldwin	9/1901	2nd rebuild
17	Norris	1902	2nd rebuild
7	Baldwin	- ? -	0-6-0 -> 2-6-0
9	Winans	- ? -	
18	Norris	- ? -	0-8-0 to 0-10-0
19	Norris	- ? -	0-8-0 to 0-10-0
20	H-B	- ? -	0-8-0 to 0-10-0
21	H-B	- ? -	0-8-0 to 0-10-0
22	Winans	-?-	2nd rebuild

Repairs and rebuildings continued into the 1940's. The shops equipment was available for charitable work as well. In 1937, the Mount Savage Fire Department built their own fire truck, with the help of the railroad shops (Cumberland Times, August 8, 1937). The engine guys at the Shops liked to relax in the evening and on the weekend, and many built their own live steam model railroads, or steam powered "parade engines."

Engine Construction at Mount Savage for the C & P

The following table shows 31 identified engine constructions at the Mount Savage Shops for the C & P. Engine 34 was not completed, and its boiler went to the Western Maryland Railway.

No.	date	wheels	notes
24	1868	0-10-0	First engine
3	1888	2-8-0	was number 18. Sold Miller's Creek
4	1889	2-8-0	was number 20. Sold Miller's Creek
11	5/1889	2-8-0	ex-51

12	9/1889	2-8-0	ex-52
13	9/1890	2-8-0	ex-53
14	5/1891	2-8-0	ex-54
15	12/1891	2-8-0	ex-55
16	8/1892	2-8-0	ex-56
7	11/1892	2-6-0	sold to 'a steel Co. in Pa'
8	1892	2-6-0	
17	1895	2-8-0	ex 57
18	1896	2-8-0	ex-58
19	8/1897	2-8-0	ex-59, sold, Millers Creek
20	6/1898	2-8-0	ex-60
21	1899	2-8-0	
22	1899	2-8-0	ex-61
26	9/1899	2-8-0	ex-62
24	9/1901	2-8-0	
9	5/1902	4-6-0	
25	12/1902	2-8-0	
10	10/1903	4-6-0	ex-30
23	10/1904	2-8-0	ex-19
32	1910	2-8-0	
28	6/1910	2-8-0	
27	2/1910	2-8-0	
29	1912	2-8-0	
30	1913	2-8-0	
31	1915	2-8-0	
33	1917	2-8-0	Last unit
34	1917	2-8-0	not completed; boiler went to Western Maryland

Millers Creek Railroad, in Van Lear Kentucky, was also a Consolidation Coal company town with many parallels with Mount Savage.

Engines built at Mount Savage for other roads

Mount Savage Shops built many locomotives for other roads. Narrow gauge equipment was built under contract to T. H. Paul, of Frostburg. Later, Mount Savage began to market their own narrow gauge equipment, built to the Paul pattern. It is probably safe to say that sales after 1883 resulted from the issuance of the catalog.

The Mount Savage Works sent two engines to the National Exposition of Railroad Appliances in Chicago in May of 1883. Both were 3 foot gauge. The 4-4-0 unit had 12"x20" cylinders, 48-inch wheels, and weighed 41,000 pounds. The 2-6-0 unit had 14"x18" cylinders, 40-inch wheels, and weighed 49,000 pounds. The disposition of the engines after the show is unknown.

Engines need a tender to carry the coal and water. These were built at Mount Savage as well. The early rule of thumb was that the ratio of water to coal consumption is 7:1. Tenders were constructed from heavy gauge sheet iron, usually 1/8" plate on the sides, and 3/16" on the bottom. Due to rust, these units had approximately a 10-year working life, but were easily repaired. The horseshoe-shape water tank was the favored design. Wooden frames were used until about 1870, when iron frames were substituted. The C & P used inside bearings on the tender trucks. C & P tenders had a water capacity of 5-6000 gallons. Rumor has it that the C & P deliberately used small tenders, because otherwise their engines were "borrowed" by the B&O to do the Brunswick run with loaded coal trains on the weekends. C & P engines left in Cumberland or Piedmont over the weekend were frequently used by the B&O to reduce wear-and-tear on their own motive power fleet.

Mount Savage Products travel far

C & P number 32 went to the Winchester & Wardensville circa 1945. This engine, a 2-8-0, was built in Mount Savage in 1910. It maintained its road number. In 1941, the W&W reorganized to the Winchester & Western. The engine worked on that line until it was scrapped in 1953.

Austin and North Western (A&NW) railroad, Austin, Texas, No. 5, 2-8-0, 3-foot gauge, 15" x 18" cylinders, 28 tons, 36" drivers. This engine was listed as Mount Savage serial No. 33 by the Southern Iron & Equipment (SI&E) Co. in 1918. It went to Tallahalla Lumber Co. It was also listed as SI&E No. 1264, before being sold to the Madrozo Sugar Company of

Cuba as their No. 4 on 26 August 1918. It was reported out of service in 1948. It may still be sitting in an overgrown shed on a Cuban Sugar plantation.

The Mount Savage Shops in the Twenty Century

Construction of locomotives ceased at Mount Savage around the time of the First World War. Heavy repair and rebuilding of locomotives continued until the time of the Second World War. The machine shops were used into the 1950's. New technologies were introduced, such as electrical lighting and electrical welding.

The Mount Savage shops, constructed with 30-inch thick stone walls, had a floor space of about 22,000 square feet. Dirt floors were preferred for the forge, blacksmith shop, and for welding. Concrete pads were poured for the machine tools at a later date. Motive power was overhead line shafts, and shop air. When a lot of machines came on line at the same time, or there was an excessive use of shop air, the power shop foreman would come running. The power shop also generated electricity, and heated the building. Until World War II spurred the development of small, lightweight electrical handheld power tools for the aircraft industry, most industrial shops used air tools. The Mount Savage shops did not have a large overhead crane capable of lifting and transferring a complete locomotive, so these operations were done manually. Before a locomotive was lifted, it was important to remember to first remove the whistle. It would be knocked off by a roof beam if it were not removed. This was confirmed several times. To un-wheel a locomotive, it would be jacked and blocked. Jacking was done with hydraulic (water) jacks. To move a locomotive in the shop, a series of pulleys, chains, and fixed floor anchors would be used with a transfer table arrangement. The shop engine served as the motive power. A rewheeling of a locomotive would be done by a 3-4 man team in one 8-hour shift.

When a locomotive came into the shops for a rebuild, the frame legs would be built up by electric welding (after that became prevalent) by a boilermaker. The frame would be squared manually, with grinding and wedges. It was important to accurately align the axles to the frame, and parallel to each other. Axles or journals rarely needed to be replaced. Before welding became popular, some parts needed to be discarded due to excessive wear. "Texas Jack", a highly respected mechanic from the Texas & Pacific Railroad, was the expert at truing frames in the C & P shop.

Brake adjustments were made from underneath, in a cramped location. The proper open-end wrench was supposed to be used, but most chose an alligator wrench. Cylinder heads used a special copper wire gasket that was hard to fit. Later, the master mechanic specified the use of an asbestos gasket, that was easier to fit, but blew out more frequently. It was highly favored by the shop crew, because of the overtime it generated for repairs.

If the operating crew were frugal with the valve oil, the rear rods of the eccentrics would sometimes bend. The engines would then limp in, and generate even more overtime for the shop crew.

Locomotive rebuilds and inspections were scheduled based on flue operating times. After a rebuild, the official operating time on the flues didn't start until a fire was built in the firebox. Thus, if the engine wasn't scheduled to go back to work immediately, it would be moved to a storage location using compressed air. The valve gear would be put in reverse, and the engine towed forward, to pump air into the boiler with the pistons. The process was repeated, until sufficient pressure was built up, and then the engine was shoved on the turntable and uncoupled. Aligned with the right stall, the engine could put itself into the stall with air pressure. This procedure was necessary because only one engine at a time would fit on the turntable.

In later years, Mr. Claus, the general manager, was a stickler for cleanliness and neatness in the shop. These concepts were not necessarily alien to a locomotive erecting, manufacturing, and repair shop. All tool boards had painted outlines for the various wrenches, there being a proper place for everything, and everything was expected to be in its place when not in use. There was a tag system for specialty tools checked out of the tool room. Standard bolts were stacked neatly with heads on alternate ends. An apprentice machinist was expected to make his own hand tools. The appренticed trades at Mount Savage were machinist and boilermaker.

From a solid grounding in mathematics and mechanical drawing at the high school level, apprentices would serve under experienced machinists to learn the details of their trade. Strangely, machinists were responsible for the superheaters, while boilermakers did the cowcatchers. Specialty parts such as springs or valves were typically purchased. Mount Savage shops produced piece parts for the war efforts of World War I and II. Gun mounts were made for the war department. Between the wars, a half day shift on Saturday was the norm, for cleaning and shop maintenance. Men with automobiles were allowed to use one of the car sheds to work on their vehicles in the afternoon.

For a rebuild, the locomotive's body-fit frame bolts would need to be removed. After the nut was loosened, the bolt would be soaked in penetrating oil, then smacked with a sledge. If it didn't budge, then a special black powder cannon with a plunger would be used to blow the bolt out. It was considered a point of honor with the shop crew not to cut or drill the bolt.

At Mount Savage, the boiler backheads were fashioned by teams of three boilermakers the old fashioned (i.e., the hard) way. Pre-punched holes in a steel plate would be fitted over pins in a forming plate, and the assembly placed over a forge. Using mallets made from gum wood, the boilermakers would manually form the plate. They hit it in a rhythm, all in the same place. The firebox door would be formed last. While white-hot, the opening would be flanged. Boilermakers were always hard of hearing.

When assemblies could not easily be done at the Mount Savage shops they would be jobbed out to larger shops. This was done, for example, for boiler rings, which were rolled in the larger capacity roller in the B&O shops in Cumberland, in later years. Holes for rivets were always punched, not drilled.

The boilermakers were artisans of their trade. On the C & P engines, the smokebox saddle was not machined, as the shops did not have a tool large enough. The structure was laid out with a wooden form and chalk, and the boilermaker formed it by hand with an air chisel. All lap-jointed, welded seams were feathered. This was also done with a hand-held air chisel. The quality of the work depended on the skills and ability of the boilermaker.

In later years, the crossheads were reamed by a tool developed by the shop. It was a series of blades on a steel mandrel, with wooden spacers. Later, ultraprecision crossheads were purchased from Timkin. These came in halves, and were clamped over the piston rods. Sellers-type injectors were used. The locomotive exterior piping was unique, and only fit one way. Although pre-bent in a pipe bender, it was not unknown for a boilermaker to use the spokes on the drive wheels and muscle power to tweak a piece of pipe to fit into a particular location.

Contributions to Canal Boats

Not content with locomotive construction, Mount Savage resident Patrick O'Conner filed U. S. Patent 315,159 on April 7, 1885 for a canal boat propeller. He assigned half to James T. O'Conner, his father. Witnesses were R. T. Semmes and Fred S. Wilson. This was essentially a driven waterwheel, centered in the boat. It is not shown, but must have been driven by a steam engine.

The author participated in an interesting archaeological hunt in Mount Savage a few years ago. I refer to this as,

The Curious Case of the Boiler in the Basement; a Mount Savage Thriller.

It started when I received a curious phone call from a friend of mine. He lives and works in Florida, but was visiting his sister in their hometown of Mount Savage. He was shown something puzzling by a mutual friend, and called me to investigate it, as he had to return the next day to work. I called the mutual friend, and set up a meeting during Mount Savage Days, 2013. This was one of those phone calls you get once in a lifetime. Names are withheld to protect the innocent.

Our mutual friend had recently bought the old 5 & 10-cent store in town. Duffy's store closed around 1968. It was built of Mount Savage enameled brick, with pressed tin ceilings. Duffy had purchased the building before World War 2. Duffy (Adolph J. Waitekunas) was the tenant of the store when the boiler was installed. The building was built in 1899, and was not yet completed when the owner, R. H. Brannon, died unexpectedly at the age of 45. The building was owned by Mathew Mullaney, and later sold to Duffy. Duffy died in 2010 at the age of 89.

When the friend went into the basement, the heating boiler seemed strange. He called in his friend (from Florida), whose jaw dropped at the sight of what appeared to be an old locomotive boiler, re-purposed to heat the building. He knew it wasn't built as a stationary boiler, because of the design. He should know, as he is restoring a steam locomotive himself. Boilers intended for mobile use have extra reinforcing, called staybolts, in the boiler to stiffen it against the bouncing and jostling of mobile use. The basement boiler had staybolts.

Why this was all extraordinarily interesting was that there are no known examples remaining of the output of the Mount Savage Locomotive Works. If this were truly an artifact of the Works, it would have to be preserved. The Historical Society was contacted.

A Mister Steven Watkins, recently deceased, saw the boiler installed "sometime before World War II". It came from behind the C&P Shops by mule, and was a "challenge" (not his choice of words) to get into the basement due to weight, and the size of the door.

The 1883 Catalog of the Mount Savage Works gives the specifications for the four Paul engines, and the standard gauge 2-8-0 wheel arrangement Consolidation type.

If we could match any of these numbers to the measurements of the boiler in the basement, we would have a good indication that it was a Mount Savage product.

Parameter	American	Logging	Mogul	Mogul Consolidation
Boiler thickness	tbd	3/8"	5/16"	3/8"
Diameter	43"	31"	43"	48"
Firebox	3/8"	3/8"	3/8"	3/8"
Firebox size	48" x 19"	41" x 22"	65" x 19"	68" x 23"
Flue sheets	7/16"	7/16"	7/16"	¾"
Number of flues	130	55	130	143
Inside diameter	2"	2"	2"	2"

Armed with this data, we headed off to the basement to take pictures and do measurements. The boiler was mounted on two brick pedestals, one under the firebox and the other under the smokebox. The smokestack was extended over to the building's chimney with galvanized pipe. An electric stoker/fan *IronFireman* assembly was controlled by a timer and steam-pressure gauge arrangement. That unit still operated. The firebox was partially lagged in asbestos. The steam dome was tied into the building's plumbing for the radiators. The firebox door was a replacement, and there was no builder plate or identifying marks on the unit.

Results of measurements on actual boiler

Boiler diameter	25"
Number of flues	31
Inside diameter	1 7/8"
Boiler length	54"
Total length	119" from boiler backhead to smokebox front

Well, this was disappointing. The small number of flues indicated that this unit was not a T. H. Paul boiler. Further research was called for. It was suggested by a Locomotive Master Mechanic with extensive knowledge of steam, that the boiler resembled that of a tractor engine, used to plow fields. What seemed to confirm this were some brackets at the front of the firebox, where the steering gear would have been attached.

As I researched this, I came across another possibility. In Amanda Pauls' book on Mount Savage, there is a picture of a parade locomotive built by Jim Lancaster. It is still in town, and resembles the boiler in the basement. Perhaps the basement unit was actually built in town after all. Not necessarily at the old C&P Shops, though.

Wrap-up

So much happened in the small Town of Mount Savage during the Industrial Revolution in the United States. It contributed greatly to the U. S.'s steel industry – perhaps it is better that was located in Pittsburgh, due to the massive amount of pollution it generated.

There is much more to research and discover. There are still things to be found in Mount Savage relating to the iron works and the locomotive manufacturing. There are discoveries to be made in old attics and basements. More and more information is showing up in Internet searches. Keep looking.

I was in town with famed Railroad photographer Bill Price, when the plaque commemorating the manufacture of the first iron rail was unveiled. I gave him a ride back to Cumberland, and we had a great idea – set up a termite inspection company, offer a free inspection, and visit every attic and basement in the Town. Unfortunately, Bill passed away before we got this done. He's out along the tracks, somewhere, watching trains go by.

Glossary

ASIN – Amazon Standard Inventory Number
Bessemer Converter – furnace to make steel from iron by removing contaminants.
Bituminous coal – an organic sedimentary rock of mostly carbon.
Blacksmith – craftsman wheo creates items from wrought iron.
Blast furnace – furnace to produce iron from iron, and using a forced blast of air to get better combustion.
Bloom – a porous mass of iron and slag, called sponge iron.
Bloomery – a crude iron production facility for small batches.
Blowing engine – a large cylinder powered by a steam engine to produce the blast for the furnace.
Bog iron – poor quality iron made from ore found in swampy areas.
Bosh - lower part of a blast furnace, between the hearth and the stack.
Chafery – a reheating hearth, to work pig iron into wrought iron.
Charcoal – produced by heating wood in the absence of oxygen. Nearly pure carbon.
Coal – a mineral, mostly carbon, with a variety of other trace elements.
Coal gas – by-product of coke production
Coke – produced by destructive distillation of coal. Is mostly pure carbon.
Finery forge – facility to produce wrought iron from pig iron, by removing carbon.
Fire clay – silica and alumina.
Flux – material to bind with and capture the impurities in the iron ore.
Forge – a heating furnace with forced draft, fueled by charcoal or coal.
Hematite – iron ore, predominate in Western Maryland. Fe_2O_3.
ICC – Interstate Commerce Commission, Federal regulatory agency.
Iron - element number 26.
ISBN – International Standard Book Number.
Limestone – sedimentary rock, calcium carbonate.
Open Hearth Furnace – converts pig iron to steel. Replacement for the Bessemer furnace.
Puddlers candle – bubbles of carbon monoxide produced in a puddling furnace. Burst and catch fire at the surface.
Pig iron – iron ore with the oxygen removed. Iron ore is mostly rust – iron

bonded with Oxygen.

PRR – Pennsylvania Railroad.

Puddling furnace – converts pig iron to steel or wrought iron. Hot air passes over the molten iron.

Reducing agent – removes oxygen from a material. Carbon is used with iron.

Reduction process – opposite of oxidation. Oxygen is removed.

Reverberatory furnace – used to make iron and mild steel. The molten iron is isolated from contact with the fuel, but does contact the combustion gases.

Rolling Mill – process to shape hot iron into long sheets by squeezing between rollers.

Siemens regenerative furnace – circa 1865 open hearth furnace design.

Slag – the impurities extracted from the iron ore; a glassy material when cooled.

Smelting - extractive technique in metallurgy to produce metal from ore.

Steel – iron with a carbon percentage of 0.2 to 2.14%. Stronger than iron.

Tuyeres – nozzle to introduce the blast into the furnace. Water-cooled.

Wrought iron – pig iron worked to reduce contaminants and carbon.

Bibliography

Aldridge, Howard Redford. "The Mount Savage Iron Works," 1929, Record of Phi Mu Fraternity, University of Maryland at College Park Libraries. Special thanks to Mrs. H. Aldridge, Frostburg, Maryland for a copy of her husband's paper on the Mount Savage Iron Works. Also see J. Alleghenies, Vol. XIV-1978.

Alexander, John H. *George's Creek Coal and Iron Company*, 1836, Baltimore, available: Frostburg State University Library, Special Collections, Call No. TN805.Z6G3. Also Pratt Library, Baltimore, call number: D9549.G4A3q.

Alexander, John Henry. *Report on the Manufacture of Iron; Addressed to the Governor of Maryland.* Annapolis: William McNeir, 1840. avail: University of Michigan Library (Publisher) ASIN-B002Y28TYG

Alexander, John H. *Contributions to a History of the Metallurgy of Iron*, Part I. 1840, published in Baltimore.

Allen, Jay Douglas. "The Mount Savage Iron Works, Mount Savage, Maryland a case study in pre-Civil War industrial development," 1970, Thesis (M.A.) - University of Maryland.

Beachley, Charles E. *History of the Consolidated Coal Company 1864-1934*, 1934, Consolidation Coal Company, New York.

Bezis-Selfa, John *Forging America: Ironworkers, Adventurers, and the Industrious Revolution*, 2003, Cornell University Press, ISBN-0801439930 .

Bowen, Mary "Mount Savage, Allegany County, Maryland" presented to the Homemakers Club of Mount Savage 1953.

Bryant, William Cullen, "Mount Savage, 1860," Saturday Evening Post,

1860, Reprinted in Tableland Trails, Vol. I, No. 3, Fall, 1953, Oakland, Maryland.

Buckley, Geoffrey L. *Extracting Appalachia: Images of Consolidation Coal Company*, 2004, Ohio University Press, ISBN 0821415557.

Carney, Charles, "The History of Mount Savage," May 1967, Project 67-014-005, Cooperative Extension Service, University of Maryland.

Clark, William Bullock. *Maryland Geological Survey, Allegany County*, 1900, Johns Hopkins Press, Baltimore, Maryland. (avail: ASIN: B000IAIH7K)

Clark, William Bullock Maryland Geological Survey: *The Limestones of Maryland.* Special publication of VIII, Part III, JHU Press, 1910, (avail. ASIN: B005HZKQLQ)

Deffenbaugh, Mrs. Roy, "History of Mount Savage, Maryland," 1968, Mount Savage High School.

Dilts, James D. *The Great Road The Building of the Baltimore and Ohio The Nation's First Railroad*, 1828-1853, 1993, Stanford University Press, ISBN-0804722358.

Francis, C. B. ; Camp, J. M. *The Making, Shaping and Treating of Steel*, 1940, 5th Ed, Carnegie-Illinois Steel Corp. , ISBN 0930767039.

Gordon, Robert B. *American Iron 1607-1900*, JHU Press, 2001, ISBN-0801868165.

Harvey, Katherine A. *The Best-Dressed Miners - Life and Labor in the Maryland Coal Region 1835-1910*, 1969, Cornell University Press, ASIB-B000RB0ZMU.

Harvey, Katherine A. "The Lonaconing Journals: The Founding of a Coal and Iron Community 1837-1840," March 1977, Transactions of the American Philosophical Society, Philadelphia, Vol. 67, Part 2.

Harvey, Katherine A. "Building a Frontier Ironworks: Problems of Transport and Supply, 1837-1840," Maryland Historical Magazine, Vol. 70, No.2, Summer 1975.

Hicks, H. Ray "The Cumberland and Pennsylvania Railroad," R&LHS Bulletin No. 66, March, 1945, pp. 36-50.

Hindle, Brooke; Lubar, Steven *Engines of Change, The American Industrial Revolution 1790-1860*, 1986, Smithsonian Institution.

Hodge, James T. *Report of the coal properties of the Cumberland coal basin,: In Maryland from surveys and examinations made during the summer of 1869*, The Major & Knapp Eng., Manufacturing and Lithography Co., avail: ASIN-B0008CHWE2

Hughes, George Wurtz, *Extracts from reports of an examination of the coal measures belonging to the Maryland mining company, in Allegany county; and of a survey for railroad from the mines to the Chesapeake and Ohio canal, at Cumberland,* 1837, Printed by Gales and Seaton, Washington. (available: Pratt Library, Baltimore).

Knowles, Anne Kelly *Making Iron, the Struggle to Modernize an American Industry 1800-1868*, 2013, U. Chicago Press, ISBN 978-0-226-44859-6.

Knowles, Anne Kelly, *Mastering Iron*, U. Chicago Press, 2013, ISBN 0-226-44861-4.

Lacoste, Kenneth C., Wall, Robert D. *An Archaeological Study of the Western Maryland Coal Region: The Historic Resources*, 1989, Maryland Geological Survey, ASIN B00073ACW6.

Lesley, J. Peter. *The Iron Manufacturer's Guide to the Furnaces, Forges, and Rolling Mills of the United States*, New York: John Wiley, 1859. (reprinted 2010, ISBN-: 1143274830).

Mellander, Deane. *Rails to the Big Vein, the Short Lines of Allegany County, Maryland,* January, 1981, Potomac Chapter, NRHS, Inc. ASIN-B000GR9LXY.

Mellander, Deane. *Cumberland and Pennsylvania Railroad,* 1981, Carstens Publishers, Inc., ISBN 911868-42-9.

Minor, D.K. (ed.), American Railroad Journal, Summer, 1844, tour of Mount Savage. Avail: http://onlinebooks.library.upenn.edu/webbin/serial?id=amrailj

Nicolls, William Jasper. *Above Ground and Below in the Georges Creek Coal Region,* 1898, Consolidation Coal Company, Baltimore, ISBN-1272115011.

Osborn, H. S. *The Metallurgy of Iron and Steel,* 1869, ASIN B00N1AVSFM.

Pangborne, J. G. *The Golden Age of the Steam Locomotive,* with over 250 illustrations, Dover Publications, 2003, ISBN 0-486-42824-9. reprint of 1894 volume by Winchell Publishing Co., NY.

Park, John R. *Maryland Mining Heritage Guide,* 2002, Stonerose Publishing Co. ISBN-0-9706697-2-0.

Paul, Amanda *Mount Savage, Images of America Series,* Arcadia Publishing, August 25, 2004, ISBN- 0738516805.

Peacey, J. G. *The Iron Blast Furnace: Theory and Practice* (Materials Science & Technology Monographs), 1979, Pergamon, ISBN-0080232582.

Pearson, Henry Greenleaf *An American Railroad Builder: John Murray Forbes,* Houghton, Mifflin & Co., 1911, ISBN-1430469684.

Randolph, B.S. "History of the Maryland Coal Region," Journal of the Alleghenies, Vol. XXIX-1993, pp. 47-62.

Rankin, Robert G., *Report on Cumberland Coal Basin*, 1855, New York: John F. Trow, Printer. Maryland Department of Geology, Mines, and Water Resources. Avail: books.google.com

Rees, D. Morgan *Mines, Mills and Furnaces*, 2008, AMGUEDDFA GENEDLAETHOL CYMRU (NATIONAL MUSEUM OF WALES), ISBN-0118800833.

Richards, William A. *Forging of Iron and Steel: A Text Book for the Use of Students in Colleges, Secondary Schools and the Shop*, 2013, Forgotten Books (reprint), ASIN-B0083IN65C .

Richards, William M. "An Experiment in Industrial Feudalism at Lonaconing, Maryland 1837-1860." M.A. Thesis, University of Maryland, 1950. Available at FSU Ort library, special collections.

Rowen, Ele *Rambles in the Path of the Steam-Horse*, Philadelphia: William Bromwell and William White, 1855, ISBN-1236549627.

Scharf, J. Thomas, *History of Western Maryland, being a history of Frederick, Montgomery, Carroll, Washington, Allegany, and Garrett Counties from the earliest period to the present day, including biographical sketches of their representative men*, 2 volumes, Philadelphia, 1882. Vol 1: ISBN-1434426297, Vol. 2: ISBN-ISBN-1434426300. (this is very handy: Long, Helen, Index to Scharf's History of Western Maryland, Volumes I and II, 2013, ISBN-0806345667.)

Schwartz, Lee G.; Feldstein, Albert; Baldwin, Joan H. *Allegany County, A Pictorial History,* 1980, The Donning Co., Virginia Beach, Virginia. ISBM-0898650178.

Singewald, Jr., Joseph Theophilus *Report on the Iron Ores of Maryland, with an Account of the Iron Industry*, Part III of Maryland Geologic Survey, Volume Nine, Baltimore: JHU Press, 1911. Reprinted 2013, ISBN-129519886X.

Stakem, Patrick H. *Cumberland & Pennsylvania Railroad Revisited*, 2002, PRRB, ISBN 0-9725966-0-7.

Stakem, Patrick H., "Coal to the Western Terminus; Canal-Railroad Connections in Cumberland, Maryland" Sept. 1995, On the Towpath, publication of the C&O Canal Historical Society, Vol. XXVII, no. 3, p. 10-11.

Stakem, Patrick H. "The Earliest Railroad Activities in Western Maryland, 1828-1870," 1996, J. Alleghenies, Vol. XXXII, ISSN-0276-7449.

Stakem, Patrick H., "The Mount Savage Rail Road 1845-1854,"June 1995, The Automatic Block, (publication of the Western Maryland Chapter, National Railway Historical Society) Vol. 17, No. 6; reprinted in Cumberland Times, Sept. 30, 1995, Railfest special section.

Stakem, Patrick H., *The History of the Industrial Revolution in Western Maryland*, 2011, PRRB Publishing, ASIN-B004LX0JB.

Stakem, Patrick H. "The Mount Savage Locomotive Shops," National Railway Historical Society (NRHS) Bulletin, Spring/Summer 1999.

Stakem, Patrick H. *Lonaconing Residency, Iron Technology and the Railroad*, 2011, PRRB Publishing, ASIN- B004L62DNQ.

Stegmaier, Harry, Jr., et al. *Allegany County - A History*, 1976, McClain Publishing, Parsons, West Virginia, ISBN-0870122576.

Stough, Bradley *The Metallurgy of Iron and Steel*, 1934, McGraw-Hill, ASIN-B00085TCWO.

Swank, James Moore *History of the Manufacture of Iron in All Ages: And Particularly in the United States from Colonial Time to 1891*, reissue 2011, Cambridge University Press, ISBN-1108026842 .

Thomas, LL.D., James W., Williams, Judge T.J.C. *History of Allegany County, Maryland*, 1923, reprinted 1969, Baltimore, Regional Publishing

Co. (p. 626-628, Millholland).

Walsh, Richard; Fox, William Lloyd *Maryland: A History: 1632 - 1974,* Maryland Historical Society, Baltimore, 1974, ASIN-B0000E921V.

Ware, Donna M., *Green Glades and Sooty Gob Piles*, 1991, Maryland Historical Trust, ISBN-1878399012.

Weld, Henry Thomas "A Report made by Henry Thomas Weld, esq., of the Maryland and New York Iron and Coal Company's lands in the county of Alleghany (sic) and State of Maryland. 1839. avail: Lehigh University Library, Bethlehem, Pennsylvania, https://library.lehigh.edu/
1.
White, John H. Jr., *A History of the American Locomotive Its Development 1830-1880*, Dover Publications, 1968; Reprinted 1979, ISBN 0-486-23818-0.

Wigginton, Eliot (ed), Foxfire 5, 1979, Anchor Books, p. 77-207. (Iron manufacturing) ISBN-0385143087.

Resources

www.mountsavagehistoricalsociety.org

Archives of Maryland, aomol.msa.maryland.gov/ (searchable)

Dictionary of American Biography, 1930, New York: Charles Scribner's Sons.

The Origin of the Iron Industry in Maryland,
avail: http://terpconnect.umd.edu/~gdouglas/ironores/pages/origin.html

"Lonaconing - Home in the Hills," 1986, Lonaconing, Maryland, avail: Allegany County Library, Georges Creek Branch, Lonaconing.

Report of President & Board of Directors of Cumberland Coal & Iron Company to the Stockholders, Feb. 11, 1853, New York: John F. Trow,

Printer.=, avail:books.google.com

Charters, Acts of Legislation, and By-laws relating to the Consolidation Coal Company and the Cumberland and Pennsylvania Railroad Company of Maryland, 1872, New York, William R. Vidal, Stationer.

Cumberland Sunday Times, March 24, 1957. (Mt. Savage)

Heritage Review, newsletter of the Preservation Society of Allegany County, Cumberland, Maryland. Various issues. Vol. 27, No. 10, Oct. 1999, article on the iron mines.

Scientific American, Vol. 6, No. 8 November 9, 1850. Manufacture of Railway Iron in Mount Savage.

New York Times, March 31, 1858, Page 2, regarding beginning operations of the rolling mill at Mount Savage.

Mayer, Brantz "June Jaunt," Harper's Monthly, April 1, 1857, (Mount Savage)

Washington Post, Sept 4, 1897 "The Lulworth Tennis Tourney, at Mount Savage, a notable society event, which continued three days, ended with the following as champions for 1897…".

Scientific American, Vol. 3, No. 9, Nov 20, 1847, "The Mount Savage Iron Works have been sold by the Sheriff for over $200000. The purchasers were Messrs. Corning and Winslow of Albany, N. Y. "

Scientific American, Sept. 11, 1847, Vol. 2, No. 51, "The Mount Savage Iron Works are to be sold on the 9th of October, under execution, at the suit of the English bond holders and others...."

www.mountsavagehistoricalsociety.org

"Opening of the Mount Savage Railroad Extended," Miners Journal, Cumberland, MD Sept 24, 1852.

Allegany County Deed, Mount Savage Iron to C & P RR, Jan 2, 1854, Ref: 12-339.

J. Franklin Institute, Vol. 38, Dec. 1844, pp 382-283, award #2705.

wikipedia, various.

Iron Production: Maryland's Industrial Past – the Iron Making Process, http://www.hmdb.org/Marker=19110.

Directory to the Iron and Steel Works of the United States, 2012, Ulan Press, (reprint) ASIN – B009C80K8I. (various volumes available. Volume 10 is dated 1889. Originally produced by the American Iron and Steel Association).

A Welsh Ironworks at the Close of the Seventeenth Century, www.genuki.org.uk/big/wal/ironworks.html

The following item reside in the Maryland Historical Society, Baltimore, MD. www.mdhs.org

Consol Coal Company records:
http://digital.library.pitt.edu/cgi-bin/f/findaid/findaid-idx?=type=simple;c=ascead;view=text;subview=outline;didno=US-PPiU-ais201103

Wikipedia, various

Patents

The following patents relate to Mount Savage activities. The full text of patents can be found at the U. S. Patent Office website, and on Google patents.

- *Improvement in the Construction and Heating of Furnaces for Metallurgic Operations*, C.E. Detmold of New York, N.Y. Number

3,176, July 1843.

- *Method of Effecting Combustion in Furnaces and Flues of Steam Boilers*, C. E. Detmold,
  NY, 1843, Number 33645A

- *Mode of Securing the Ends of Railway Bars*, C.E. Detmold, 22,168, Nov. 1858.
- *Rolling Railway-Bars*, John W. Brown, Savage Iron Works, Allegany County, Maryland, 14552, April 1856.

- *Improved Machine for Compressing Puddle-balls*, John F. Winslow, 34177, 1862.

- *Malleable Iron from Ores*, John F. Winslow, 4526, 1846, Troy, NY.

- *Rolling Puddlers Bars into Blooms*, John F. Winslow, Troy, N.Y., Number 5660, July 1848.

- *Canal Boat Propeller,* Patrick O'Conner, April 7, 1885.

Glossary

ASIN – Amazon Standard Inventory Number.
Bituminous coal – an organic sedimentary rock of mostly carbon.
Blacksmith – creates items from wrought iron.
Blast furnace – furnace to produce iron from iron, and using a forced blast of air to get better combustion.
Bloom – a porous mass of iron and slag, called sponge iron.
Bloomery – a crude iron production facility for small batches.
Blowing engine – a large cylinder powered by a steam engine to produce the blast for the furnace.
Bog iron – poor quality iron made from ore found in swampy areas.
Bosh - lower part of a blast furnace, between the hearth and the stack. Hottest place in the furnace
Chafery – a reheating hearth, to work pig iron into wrought iron.
Charcoal – produced by heating wood in the absence of oxygen. Nearly pure carbon.
Coal – a mineral, mostly carbon, with a variety of other trace elements.
Coal gas – by-product of coke production
Coke – produced by destructive distillation of coal. Is mostly pure carbon.
Finery forge – facility to produce wrought iron from pig iron, by removing carbon.
Fire clay – silica and alumina.
Flux – material to bind with and capture the impurities in the iron ore.
Forge – a heating furnace with forced draft, fueled by charcoal or coal.
Hematite – iron ore, predominate in Western Maryland
Iron - element number 26.
ISBN – international standard book number.
Limestone – sedimentary rock, calcium carbonate.
Open Hearth Furnace – converts pig iron to steel. Replacement for the Bessemer furnace.
Pig iron – iron ore with the oxygen removed. Iron ore is rust – iron bonded with Oxygen.
Puddlers candle – bubbles of carbon monoxide produced in a puddling furnace. Burst and catch fire at the surface.
Puddling furnace – converts pig iron to steel or wrought iron. Hot air

passes over the molten iron.

Pyrolysis – heating in the absence of oxygen to drive off contaminants

Reducing agent – removes oxygen from a material. Carbon monoxide is used with iron.

Reduction process – opposite of oxidation. Oxygen is removed.

Reverberatory furnace – used to make iron and mild steel. The molten iron is isolated from contact with the fuel, but does contact the combustion gases.

Rolling Mill – process to shape hot iron into long sheets or bars by squeezing between rollers.

Siemens regenerative furnace – circa 1865 open hearth furnace design.

Slag – the impurities extracted from the iron ore; a glassy material when cooled.

Smelting - extraction technique in metallurgy to produce metal from ore.

Steel – iron with a carbon percentage of 0.2 to 2.14%. Stronger than iron.

Tuyeres – nozzles to introduce the blast into the furnace. Usually, Water-cooled.

Wrought iron – pig iron worked to reduce contaminants and carbon.

My WM/History books

Stakem, Patrick H. *Cumberland & Pennsylvania Railroad Revisited*, 2011, PRRB Publishing, not currently in print

Stakem, Patrick H. *Eckhart Mines, The National Road, and the Eckhart Railroad*, 2011, PRRB Publishing, ASIN B004KSQVWO.

Stakem, Patrick H. *The History of the Industrial Revolution in Western Maryland*, 2011, PRRB Publishing, ASIN B004LX0JB2.

Stakem, Patrick H. *Down the 'crick: the Georges Creek Valley of Western Maryland*, 2014, PRRB Publishing, ASIN B00LDT94UY.

Stakem, Patrick H. *Lonaconing Residency, Iron Technology & the Railroad*, 2011, PRRB Publishing, ASIN B004L62DNQ.

Stakem, Patrick H. *T. H. Paul & J. A. Millhollland: Master Locomotive Builders of Western Maryland*, 2011, PRRB Publishing, ASIN B004LGT00U.

Stakem, Patrick H. *Tracks along the Ditch, Relationships between the C&O Canal and the Railroads*, 2012, PRRB Publishing, ASIN B008LB6VKI.

Stakem, Patrick H. *From the Iron Horse's Mouth: an Updated Roster from Ross Winans' Memorandum of Engines*, 2011, PRRB Publishing, ASIN B005GM4O12.

Stakem, Patrick H. *Iron Manufacturing in 19th Century Western Maryland*, 2015, PRRB Publishing, ASIN B00SNM5EIU.

Stakem, Patrick H. *Railroading around Cumberland*, 2012, Arcadia Press, ISBN- 0738553654.

Stakem, Patrick H. *Cumberland (Then and Now)*, 2012, Arcadia Press, ISBN-0738586986 , ASIN B009460QNM.

Stakem, Patrick H. *Fort Cumberland, Global War in the Appalachians: a Resource Guide*, 2012, PRRB Publishing, ASIN- B0088BWK06.

Stakem, Patrick H. *Ross Winans, an ingenious mechanic of Baltimore,* 2017, PRRB Publishing, ASIN- 1520207298.

Stakem, Patrick H. *Mount Savage, Iron Empire,* 2016, ISBN-978-1549650413.

Stakem, Patrick H. *Studebaker Wagons,* 2018, ISBN-*978-109146490* .

Stakem, Patrick H. *Riverine Ironclads, Submarines, and Aircraft Carriers of the American Civil War,* 2019, ISBN-978-1089379287.

Stakem, Patrick H. *The Snowdens' Iron Works,* 2019, ISBN-978-1070945699 .

Stakem, Patrick H. *Transportation Options on the Frontier,* 2019, ISBN-978-1091059481.

Stakem, Patrick H. *Savage Factory, Cotton to Canvas, by Water & Steam,* 2019, ISBN-978-1731437983.

www.ingramcontent.com/pod-product-compliance
Lightning Source LLC
Chambersburg PA
CBHW020929180526
45163CB00007B/2945